2

PROGRESS IN NEURAL PROCESSING

Series Advisors

Alan Murray *(University of Edinburgh)*
Lionel Tarassenko *(University of Oxford)*
John Taylor *(King's College, London)*
Andreas Weigend *(University of Colorado)*

Parallel Implementations of Backpropagation Neural Networks on Transputers

A Study of Training Set Parallelism

Parallel Implementations of Backpropagation Neural Networks on Transputers

A Study of Training Set Parallelism

P. Saratchandran
N. Sundararajan
Shou King Foo

School of Electrical & Electronic Engineering
Nanyang Technological University
Singapore

World Scientific
Singapore • New Jersey • London • Hong Kong

Published by

World Scientific Publishing Co. Pte. Ltd.

P O Box 128, Farrer Road, Singapore 912805

USA office: Suite 1B, 1060 Main Street, River Edge, NJ 07661

UK office: 57 Shelton Street, Covent Garden, London WC2H 9HE

Library of Congress Cataloging-in-Publication Data
Parallel implementations of backpropagation neural networks on
    transputers : a study of training set parallelism / editors. P. Saratchandran,
    N. Sundararajan, Shou King Foo.
        p.   cm. -- (Progress in neural processing : 3)
        Includes bibliographical references and index.
        ISBN 9810226543
        1. Parallel processing (Electronic computers)   2. Neural networks
    (Computer science)   3. Back propagation (Artificial intelligence)
    4. Transputers.   I. Saratchandran, P.   II. Sundararajan, N.
    III. Foo, Shou King.   IV. Series
    QA76.58.P3773   1996
    006.3--dc20
                                                            96-12119
                                                                CIP

British Library Cataloguing-in-Publication Data
A catalogue record for this book is available from the British Library.

Printed in Singapore.

To
my parents
my wife Jaya
and my children Para and Hemanth

                    P. Saratchandran

To
my grandparents
my wife Saraswathi
and my daughters Sowmya and Priya

                    N. Sundararajan

To
my parents

                    Shou King Foo

Preface

Purpose and Goals

The aim of this book is to present a detailed analysis of training-set parallelism for Multilayer Feedforward Neural Networks in which backpropagation is (BP) is used as the training algorithm. The analysis is motivated from the realization that with a good mathematical model of the parallelized Multilayer Perceptrons (MLP) running on a parallel hardware, it should be possible to theoretically predict the time for a training epoch without the need to carry out any hardware implementations. Such a model will be a very valuable tool in cost benefit analysis in which the benefit of adding extra processors has to be weighed against the cost of these processors. A good mathematical model is also useful in finding answers to questions like what is the optimal allocation of patterns for a given processor architecture, or what is the optimal number of processors for a given pattern allocation strategy. These are non-trivial questions and have been addressed and answered in the various chapters of the book.

As the title of the book indicates a network of transputers is used as the hardware platform on which to run the parallelized MLPs. Although the transputer network is used as the hardware platform the principles behind the mathematical analysis and the model that we have developed can be used for any message passing multiprocessor network (eg. DSP networks) and not just transputer networks. We have used transputer network in this study as we have the necessary hardware in our laboratory and we can check the theoretical models developed herein straight away.

The configuration of transputers in the processor network is in the form of a pipelined ring. Thus the mathematical model is for the parallelized MLP running on a pipelined ring transputer network. The model is developed by first decomposing the various operations of the parallelized BP algorithm into computing and communication operations and then expressing these in terms of elemental computing and communication operations of the transputers. An expression to predict the time for a training epoch is then derived in terms of the timings for these elemental operations. The expression is also used to theoretically find the optimal allocation of patterns that would minimize the training time for a given homogeneous transputer network.

After validating the model for homogeneous processor networks, extension of the model to cover heterogeneous transputer networks i.e. a processor network consisting of transputers of different speed and memory is carried out. Motivation for this extension is from the fact that in most laboratory inventory on transputers there will be some old and some new models of the processor and using all the available processors efficiently it should be possible to obtain higher performance than using processors of a single model. However the problem of theoretically finding the optimal allocation of training patterns for a heterogeneous processor network is a lot more complicated than the homogeneous processor case. The analysis for heterogeneous transputer network and their experimental verification provides a valuable tool for optimal use of all the available transputers in a laboratory in the context of training neural networks.

Finding the optimal mapping of a neural network boils down to solving a mixed integer optimization problem which is known to be NP complete and time consuming. Some fast suboptimal methods to the mapping problem including a method using the genetic algorithm are described in the final chapter of the book.

The book is primarily aimed at graduate students and research scientists working in the area of artificial neural networks and parallel computing. However the mathematical models developed are lucid and can be easily used by the practicing engineers for cost benefit analysis on parallel implementations.

Overview

Chapter 1 contains an overview of the BP algorithm and its implementations on some well known parallel machines.

Chapter 2 introduces some of the common topologies for transputer networks and their characteristics. The pipeline ring topology chosen for implementation and analysis in the rest of the book is then described. The chapter also describes the parallel programming language and the benchmark problems used throughout the book for model validation and extension studies.

In Chapter 3 development of the mathematical model for homogeneous transputer networks in the pipeline ring configuration is described. An expression for the time for a training epoch is derived for any given arbitrary pattern allocation in terms of the elemental timings for the processors. This chapter also provides experimental validation of the theoretical expression.

Chapter 4 considers the case of equal distribution of patterns among the processors. A simplified expression for the time for an epoch is derived for this case from the more general expression of Chapter 3 and validated through experimental results. The chapter also addresses the question of finding the optimal number of processors for a given neural network when equal distribution is used as the pattern allocation strategy. A discussion on cost benefit analysis is also carried out.

Chapter 5 presents the optimal solution to the pattern distribution problem for homogeneous processor networks. Finding the optimal distribution of patterns to minimize the time for a training epoch for a given homogeneous processor network is analyzed and shown to boil down to solving a mixed integer programming problem. Optimal distributions for several benchmark problems are obtained by solving the appropriate mixed integer programming problem. The chapter also provides a comparison of training times between optimal and equal pattern allocation schemes.

The analysis of Chapter 5 is extended in Chapter 6 to cover heterogeneous processor networks. Optimal distributions of patterns are found for heterogeneous pipelined ring transputer networks and the corresponding epoch times are obtained experimentally from laboratory built heterogeneous transputer networks. The chapter also provides a comparison of epoch times between the optimal and several random distributions generated using Monte Carlo methods.

Chapter 7 examines some suboptimal methods to the pattern distribution problem that are fast and are capable of producing results close to optimal epoch times. This chapter also shows the formulation and solution of the optimal distribution problem using genetic algorithms.

Acknowledgements

Undertaking and completing a project like this would not have been possible without the encouragement and support of many individuals. First and foremost we wish to thank Dr. Cham Tao Soon, President, Nanyang Technological University, Singapore, for providing an excellent academic environment at NTU which is a necessary prerequisite for an endeavor such as this to succeed.

We are grateful to Professor Brian Lee, Dean of the School of Electrical and Electronic Engineering, and Prof. Er Meng Hwa, Vice Dean (Academic) and Director, Center for Signal Processing, Nanyang Technological University for their constant encouragement and support during this study.

We also wish to thank Dr. Soh Cheong Boon, Vice Dean (Research) and Dr. Soh Yeng Chai, Head, Division of Control & Instrumentation Engineering for their support.

Special thanks are due to Professor Robert Newcomb, Director, Microsystems Laboratory, Electrical Engineering Department, University of Maryland, USA, for his constant advice, comments and criticisms of the earlier versions of this book. We owe a debt of gratitude to the anonymous reviewers who provided valuable comments most of which have been incorporated in the book.

Finally we extend our thanks to Mr. Richard Lim, Scientific Editor, World Scientific Co Ltd, London, UK, for extending his full cooperation and support in this effort.

Singapore   P.S
Dec. 1995   N.S
                                             S.K.F

Contents

Chapter 1

Introduction

The field of artificial neural networks is currently enjoying a surge of interest and activity as demonstrated by the number of journals and conferences devoted to this field. Several successful application of neural networks have been reported in such diverse areas like control engineering [1], pattern recognition [2], financial forecasting [3], telecommunications [4] and medical diagnosis [5]. There are many architectures and learning paradigms for neural networks but the most popular of all is the Multilayer Feedforward architecture in which the Backpropagation (BP) [6, 7] is used as the learning algorithm.

Although very popular, the backpropagation algorithm is very slow to converge and training times of the order of days and even weeks are not uncommon [8,9]. One approach to increase the speed of convergence is to modify the basic BP algorithm and this has been proposed by several authors [10–14]. Although the modifications usually increase the computational complexity of the algorithm, they reduce sufficiently the number of iterations required to reach convergence and thereby shorten the overall training time. Another approach to increase the speed of convergence is to use parallelism. Motivation for parallelism comes from the realization that the large training time of BP stems from the tremendous amount of floating point operations needed in each iteration of the algorithm and by performing many of these operations in parallel, significant reduction in the training time can be achieved. This has encouraged several researchers to study parallel implementation of the backpropagation algorithm as a means to reduce the training time [15–19].

1.1 Multilayer Feedforward Neural Networks

Multilayer Feedforward Neural Networks are layered networks as shown in Fig. (1.1) consisting of an input layer, several hidden layers and an output layer. Each layer consists of several neurons or nodes. The neurons in the input layer are mere fan-out units; no processing takes place in these neurons. The output of neurons in each layer are distributed to all the neurons in the next layer. The output of neurons of the output layer represent the network output. The output of a single neuron is a nonlinear function of the weighted inputs to that neuron. The weights represent the strength of connection between neurons. The most commonly used nonlinear neuron activation function is the sigmoidal function of the form $f(x) = \frac{1}{1+e^{-x}}$ shown in Fig. (1.2).

In the next section, we give a brief description of the basic BP algorithm and illustrate its computational complexity.

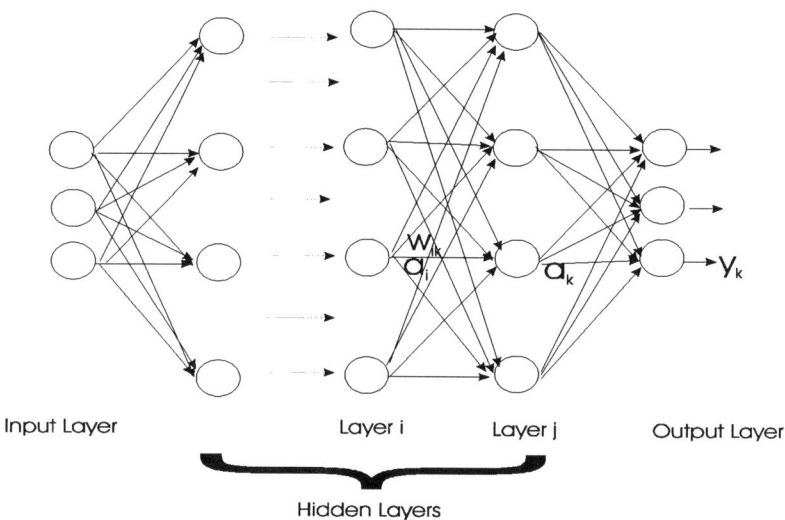

Figure 1.1: A Multilayer Feedforward Neural Network.

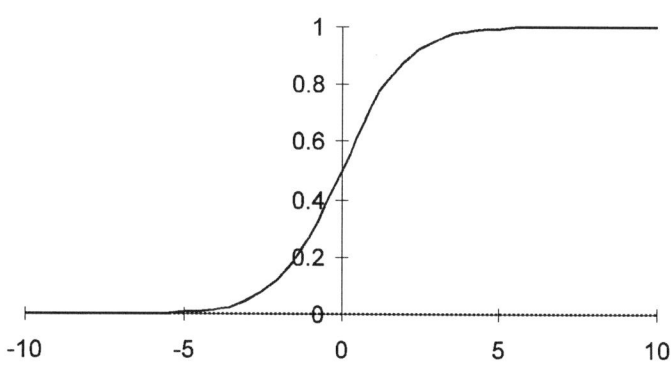

Figure 1.2: Sigmoid Activation Function.

1.2   The Basic BP Algorithm

Backpropagation is a steepest descent algorithm aimed at minimizing the squared error between the actual and the desired output for a neural network over the entire training set. The squared error between the actual network output and the desired network output for a training pattern p is defined as

$$E_p = \frac{1}{2} \sum_k (d_k - y_k)^2$$

where E_p is the network error for the pattern p, y_k and d_k are the actual and desired outputs of neuron k in the output layer and k indexes over all the output neurons. The overall error, E_{tot}, for the entire training set is the accumulated error of all the patterns in the training set.

$$E_{tot} = \sum_p E_p.$$

The algorithm consists of two phases known as the forward pass and the backward pass. During the forward pass, the activations of the neurons in each layer is calculated from the first hidden layer to the output layer of network. The activation of an individual neuron k in layer j is a non-linear function of its total input coming from the previous layer i.

$$a_k = f\left(\sum_i w_{ik} a_i\right)$$

where a_k is the activation of neuron k, w_{ik} is the synaptic weight of the connection between neuron i and neuron k. a_i is the neuron activations in layer i (See Fig. (1.1)). The nonlinear function $f(x)$ is usually a sigmoid of the form,

$$f(x) = \frac{1}{1 + e^{-x}}.$$

The backward pass of the algorithm involves calculation of the partial derivative δ_k of the error associated with each neuron k. For neurons in the output layer δ_k is given by the following equation.

$$\delta_k = (d_k - y_k)\, y_k\, (1 - y_k).$$

and for neurons in the hidden layers δ_k is given by

$$\delta_k = y_k\, (1 - y_k) \sum_l \delta_l w_{kl}$$

where l indexes over all the neurons of the next layer. After calculation of the $\delta's$ the increments $\Delta w's$ to the weights are computed as,

$$\Delta w_{kl} = \alpha\, \delta_l\, y_k$$

where α is the learning rate. In the *batch learning* mode the increments Δw's are calculated for each training pattern and accumulated. Updating of the weights in the network is performed after presenting all the training patterns in the training set. The updated value for the weight w_{kl} is given by,

$$w_{kl}(n + 1) = w_{kl}(n) + \Delta w_{kl}(n) + \beta \Delta w_{kl}(n - 1)$$

where n is the current iteration and β the momentum rate. If the *on-line learning* mode is used then the weights are updated after processing each pattern rather than at the end of the whole training set.

It is easy to see the huge amount of floating point operations needed by the BP algorithm. For a feedforward network with a single hidden layer, the floating point operations required for one forward and backward pass is approximately $2N_h(2N_l + N_1)$ multiplications and additions and $(N_l + N_h)$ sigmoid calculations, where N_1, N_l, and N_h are the number of neurons in the input, output and hidden layers respectively. To see the magnitude of the number of floating point operations needed, consider a reasonably small network of 50 input, hidden and output neurons and a training set of 1000 patterns. This network requires approximately 15 million multiplications and additions and 100,000 sigmoid calculations[1] just for one iteration (epoch) of batch learning. And what is more, the above figure of 15 million does not include the weight update calculations!

Fortunately many of the floating point operations are independent and can be performed in parallel, and this has provided much motivation for parallel implementation of the BP algorithm.

------

[1] We found each sigmoid calculation to be approximately equal to 14 multiplication operations in the transputer.

1.3 Parallelism in the BP algorithm

The backpropagation algorithm reveals several degrees of parallelism within it such as the weight parallelism, node parallelism, layer parallelism, network parallelism and training parallelism. These are explained in detail in Singer [20]. Efficient mapping of the algorithm onto a parallel hardware requires proper exploitation of these parallelism. In this regard there are mainly two paradigms known as the Network-based parallelism and the Training-set parallelism. Training-set parallelism has less communication overhead but uses batch learning which may lead to slower overall convergence for some problems. Network-based parallelism on the other hand introduces more communication overhead [21] but uses online learning which results in faster convergence especially for large training set that possess redundant information. Further it is the only choice if all the training examples are not available at the start of the training and continuous adaptation to a stream of training patterns is required. Batch learning although slow in overall convergence is, in general, the method of choice for many applications especially where high precision mapping is required [22]. A detailed discussion of the advantages and disadvantages of batch and on-line learning is given in [23].

1.3.1 Network-based Parallelism

According to this paradigm the neural network is partitioned amongst the processors so that each processor would simulate a portion of the network over the whole training set. There are essentially two approaches to partitioning a network. One is partitioning the algebraic operations performed by the network during the forward and backward passes of the BP algorithm and the other is partitioning the topology of the network. Network-based parallelism can be used for both training and recognition phase.

Algebraic Partitioning of Network:
This approach makes use of the fact that bulk of the computations in the forward and backward pass of the BP algorithm can be expressed as algebraic operations on vectors and matrices [24]. These operations can be represented in the form of a directed graph [25] which can then be mapped on to an array of processing elements. This approach provides a fine grain parallelism and is well suited for implementation on systolic hardware [25, 26].

Topological Partitioning of the Network:
In topological partitioning the neural network is 'sliced' and distributed among the processors each of which then simulates one slice of the network for the whole training set. The most popular way of partitioning the network is the vertical slicing method [15, 27–29] in which each processor gets a subset of neurons from all the layers as illustrated in Fig. (1.3). This approach has a medium grained parallelism which means it is well suited for implementation on parallel machines built using medium to coarse grain processors like transputers and digital signal processors (DSPs).

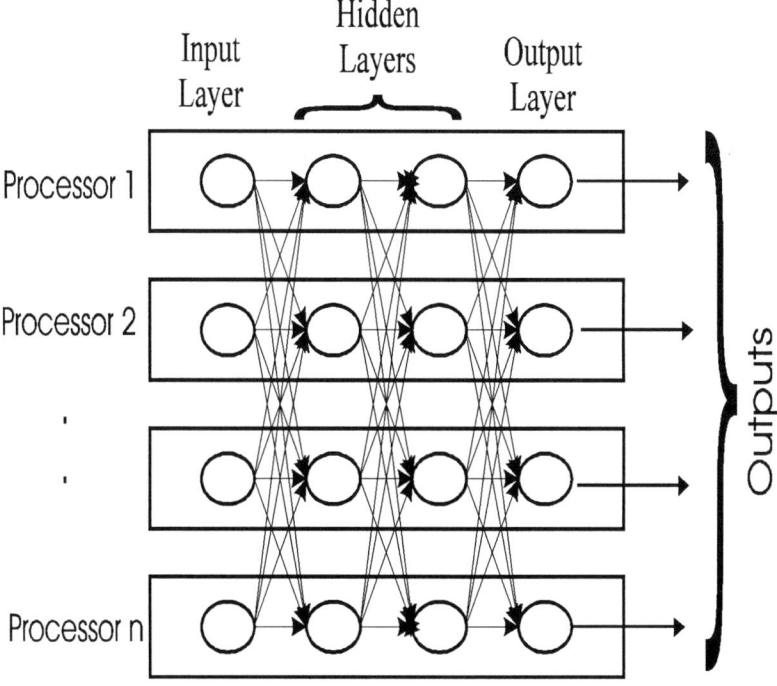

Figure 1.3: Network Parallelism.

1.3.2 Training-Set Parallelism

Training-set parallelism involves distributing the training examples among the processors, i.e., slice the training set and assign one slice to each processor while keeping a complete copy of the whole neural network in each processor node as shown in Fig. (1.4). The only communication required for this method occurs when the weights are updated. The crucial issue in training-set parallelism is how to distribute the training-set optimally so that the time for a training epoch is minimum [30, 31]. This issue is addressed in detail in

Chapters 5 and 6. Training-set parallelism can also be used in the recognition phase of BP as explained in [19].

This approach has a coarse grained parallelism and so is suitable for implementation, as in the case of network-based parallelism, on most of the commonly available Multiple Instructions on Multiple Data (MIMD) machines [32].

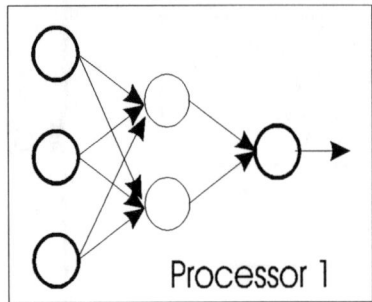 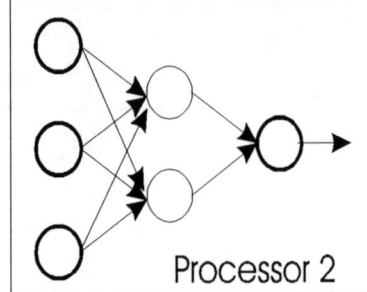

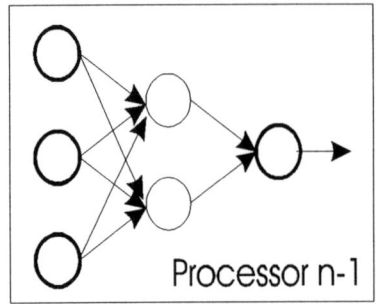 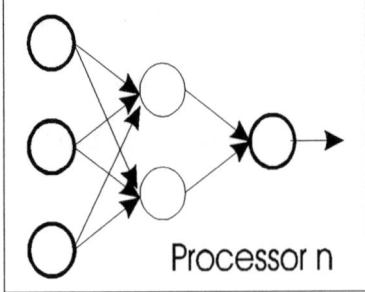

Figure 1.4: Training Set Parallelism.

1.4 Some Parallel Implementations

In this section we briefly describe some of the well known parallel implementations of the BP algorithm reported in the literature. A more detailed description can be found in [33].

WARP Systolic Array:
Pomerleau [34] has implemented the backpropagation on the WARP computer. The WARP is a programmable systolic array computer developed at Carnegie Mellon University in 1986. The WARP array is composed of a linear array of processors called "cells". Each cell has a peak computation rate of 10 MFLOPS giving a peak performance of 100 MFLOPS for the whole machine. Each cell is equipped with 32K words of fast access data memory which can be programmed separately. The initial implementation partitioned the neural network, placing one tenth of the nodes in every layer onto each processor. The improved version places the entire network onto each cell while the training patterns are being distributed among the cells. A performance of approximately 17 million connections per second is achieved.

Connection Machine CM-5:
The Connection Machine CM-5 is a massively parallel MIMD processor and uses a fat tree topology for communications. Each processing element comprise a Sparc-2 floating point processor and optional vector units. This machine was used by Liu and Wilcox [35] to implement the BP algorithm using network-based parallelism. They reported a performance of 76 million connection updates per second (MCUPS) for a 256-256-131072 neural network.

Sandy/8:

The Fujitsu Sandy/8 is a highly specialized 256 processor parallel neurocomputer for BP learning of neural networks in which the processors are connected in a ring architecture with efficient communication capabilities. Yoshizawa, Kato et al [36,37] implemented the BP algorithm on this machine using the network-based parallelizing paradigm and report a performance of 583 MCUPS.

GF11:

The IBM GF11 is a large Single Instructions on Multiple Data (SIMD) machine with up to 566 processors interconnected through a Benes [38] network. Witbrock and Zagha [39] implemented BP on this machine with 356 processors. They used the training set parallelism approach with the neural network being duplicated on each processing unit. A performance of 900 MCUPS have been reported.

MUSIC:

Multiprocessor System with Intelligent Communication (MUSIC) is a MIMD machine in which each processing element consists of a Motorola 96002 digital signal processor (DSP), 8 Mbytes of dynamic RAM, 1 Mbytes of static RAM and a programmable gate array as a communication controller. Muller, Gunzinger and Guggenbuhl [17] implemented the BP on this machine using the network-based parallelizing paradigm. A performance of 330 MCUPS is reported using sixty three processing elements.

DREAM:

Dynamically   Reconfigurable   Extended   Array   Multiprocessor

(DREAM) is a SIMD machine designed specifically for neural network implementations. The processing elements in DREAM are arranged on a 2D lattice where each processing element is connected to eight of its closest neighbors through four programmable switches. Shams and Gaudiot [18] have implemented BP on the DREAM machine and have reported a performance of 512 million connections per second.

TLA:

The Sharp Toroidal Lattice Architecture (TLA) neurocomputer is an example of a MIMD machine built using transputers. It has sixteen T805 transputers connected in a toroidal architecture and an extra transputer acts as the root processor interfacing with the host IBM PC. Fujimoto, Fukuda and Akabane [40] have implemented the BP on this machine using the network-based parallelism approach and have reported a performance of 600 KCUPS.

To summarize, in this chapter, the basic elements of the Back-propagation algorithm and the two major parallelizing paradigms have been reviewed. A brief overview of different parallel implementations of feedforward neural networks is also given. Since this study is performed specifically on transputer arrays, the hardware and software for the transputer systems are reviewed in the next chapter.

Chapter 2

Transputer Topologies for Parallel Implementation

A brief summary of various parallel implementations of backpropagation on different hardware platform was described in the previous chapter. Since we have chosen the transputers as the basic processing element in our studies, it is important to understand the basic hardware, software and topologies of transputer networks and this is the main focus of this chapter. Also, to validate the mathematical model developed, a number of well known benchmarks have been used. Since these problems appear throughout the book, we have described them in detail in this chapter.

2.1   The Transputer

The transputer is a single chip VLSI device which can be used as a multicomputer building block and is produced by the the INMOS division of SGS-Thomson. Each transputer chip consists of a processor, a memory and a set of communication links.

The main feature of the transputer is that it is a computing device in which communications are primitive operations and parallelism is a primitive construct. Several transputers can easily be connected to form a powerful processor network. A transputer network is a MIMD machine because each processor is separately programmable. The T805 family of transputer which we have used in the theoretical modeling and experiments in this book, has 4 KBytes of fast on-chip Static RAM, a 32 bit integer CPU and a 32 bit IEEE floating point unit(FPU). The CPU and the FPU can work fully in parallel. Further there are four link interfaces. These links are direct-memory-access (DMA) controlled, bidirectional serial transmission channels which can provide up to 9 MBytes of I/O throughput for a transputer.

The latest addition to the transputer family is the T9000 transputer. It consists of a super scalar 32 bit CPU with a 64 bit FPU and a communication processor together with four communication links. It has 16 KBytes of on-chip memory and an external memory interface. The 50 MHz version is said to be capable of running [41] at about ten times the speed of a T805-20 transputer. Experience on T9000 from the industry has so far been mixed.

More details on transputer family can be found in any one of several excellent books on transputers [42–44].

2.2   Topologies

Transputers can be configured in a variety of topologies so long as the incoming and outgoing connections to each processor do not exceed the number of links (i.e. four). Some of the popular topologies for transputers are ring, torus (two dimensional ring), hypercube [1], and mesh. Fig. (2.1) shows these topologies. A brief description of these are given in the following sections.

2.2.1 Ring

Ring is the simplest and perhaps the most widely used [45, chapter 6], [19,27,46] topology with transputers. Each processor in the ring network has a direct communication link to two other processors. Although simple to implement, the bandwidth of ring network is limited due its large diameter[2]. The most common forms of communication schemes and the number of steps required for these in an n processor ring are given in Table (2.1).

Table 2.1: Performance of different communication schemes for Ring topology.

Type of Scheme	Number of steps needed
one-to-one	$\lfloor \frac{n}{2} \rfloor$
one-to-all	$\lceil \frac{n}{2} \rceil$
all-to-all	$(n\text{-}1)$

---

[1]Because there are only four links only up to four dimensional hypercubes are possible.

[2]Diameter is the maximum distance in terms of number of links between any two processors in the network.

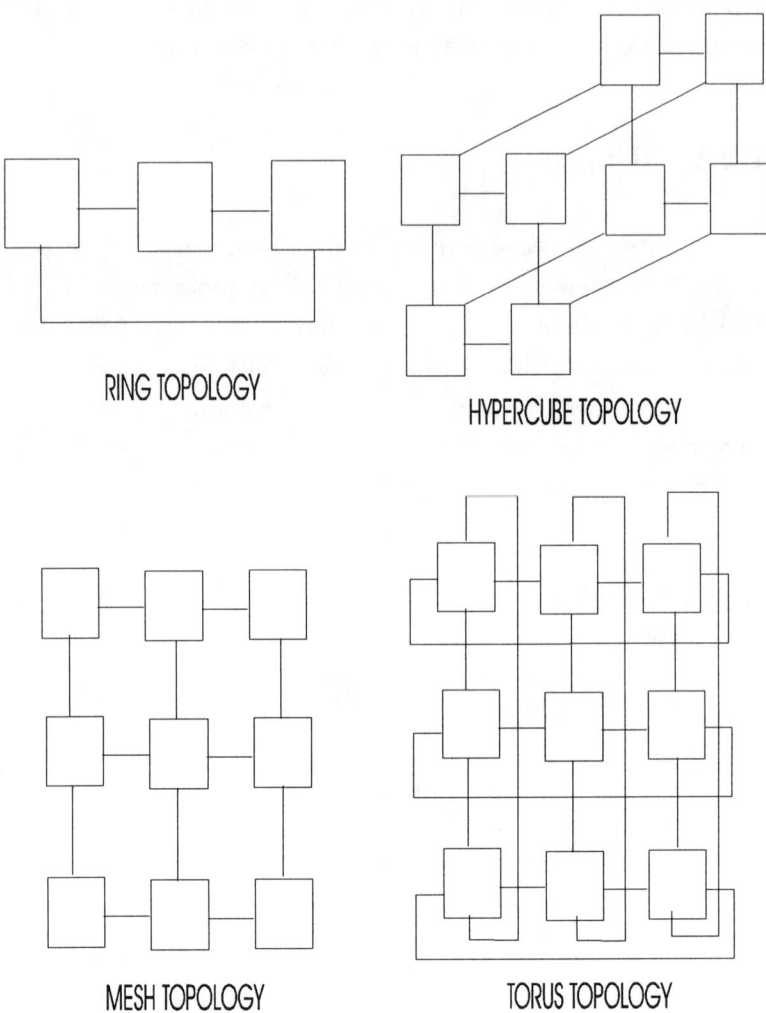

Figure 2.1: Different topologies for Transputer networks.

2.2.2   Mesh

A mesh is an extension of a linear array to higher dimensions. A two dimensional mesh has each processor directly connected to four other processors. If the number of processors in each dimension is the same then it is called a square mesh otherwise it is a rectangular mesh. Since transputers have four links, mesh is a popular topology for transputer networks. It has a diameter of $2 \left(\sqrt{n} - 1\right)$ which is smaller than that of a ring. The only drawback of mesh is probably the lack of complete symmetry caused by the nodes at the border.

2.2.3 Torus

A torus topology results from folding horizontally and vertically a mesh network and thereby obtain a completely symmetrical structure which is lacking in the mesh. The symmetrical structure also results in a smaller diameter for the torus network than that for the mesh. The diameter for a torus is $2 \lfloor \frac{\sqrt{n}}{2} \rfloor$. However folding the mesh also results in reduced scalability for the torus due to the need for wrap around links which are roughly proportional to the square root of the number of processors [42].However, torus topology is used for many transputer networks [16, 40] because of the smaller diameter.

2.2.4 Hypercube

A hypercube is a multidimensional mesh with exactly two processors in each dimension. A d dimensional hypercube would thus consists of 2^d processors. An elegant property of hypercubes is their recursive structure. A one dimensional hypercube can be constructed by connecting two zero dimensional hypercubes and so on. Such a property makes them quite attractive for designing a very wide class of algorithms. The diameter for a hypercube is smaller than that of a torus. A hypercube of n processors has a diameter of only $\log_2$ n. A possible drawback of hypercubes, as interconnection topology for multicomputers, is that they are less scalable than the torus. For instance in a d cube each of the 2^d nodes is directly connected to other d nodes through point to point links. This means that the cost of interconnection per node increases with number of nodes. In torus for example each node is connected to four other nodes no matter how many nodes are there in the processor network.

In the case of transputers, since there are only four links, hypercubes using transputers can only have up to 16 processors. This is a severe restriction and is the main reason for the lack of parallel machines built with transputers using the hypercube technique.

2.3   Topology Chosen in this study

The topology for the processor network used in this study is the ring. However the ring topology used in this book differs from the basic ring in that we have used a bidirectional ring. In terms of the communication scheme, we used a series of one-to-one communications instead of the all-to-all broadcasts [27] commonly employed in the basic ring. We have called our topology as a pipelined ring topology. The connection of transputers in the pipelined ring topology is as shown in Fig. (2.2). It can be shown theoretically that when the training-set is optimally distributed among a homogeneous array of processors connected in the pipelined ring topology, shorter epoch times will be produced as compared to the basic ring topology. Appendix A gives the proof for this. The details of the computation and communication operations performed in the pipelined ring topology is described below.

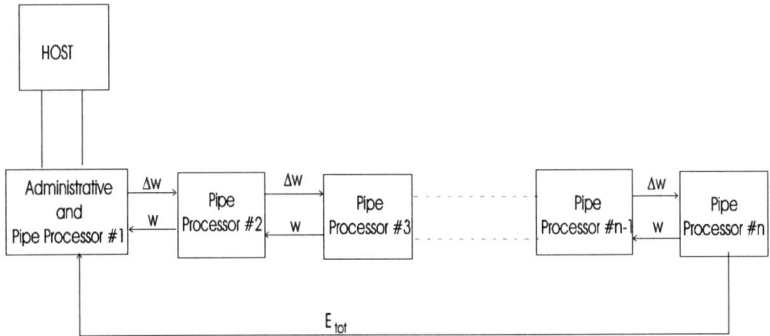

Figure 2.2: Pipelined ring connection of transputers.

In the pipelined ring topology there is one root or administrative processor and a number of pipe processors. The root processor is connected to a host machine (a 486 personal computer). Each pipe processor holds a copy of the whole neural network and the training data assigned to it. At each update cycle a pipe processor receives partial sums of weight change and error from its immediate upstream neighbor, adds to this the weight change (Δw) and error it has calculated for the training data it has been assigned, and sends the updated partial sums of weight change and error to its next downstream neighbor. The last processor in the pipelined ring will thus have the accumulated weight change and error due to all other processors in the pipeline. It adds its own weight change and error to this, and obtains the total weight change and error (E_{tot}) for the epoch. It then sends the overall error to the root processor for checking of the stopping criterion. As the root processor checks whether or not the stopping criterion is met, the last pipe processor updates the weights by adding the total weight change to the existing weight values and propagates the new weight values upstream. Each pipe processor thus receives new weights from its downstream neighbor and passes these to its upstream neighbor. The new weights will be used for the next iteration[3].

2.4   Software Used

The programming language used for all simulations in the book is the 3L Parallel C Version 2.2.2 [47]. It is a form of C language that supports the ANSI C standard. Moreover additional function calls are available that maps parallel processes onto one or more transputers.

A parallel C application consists of a collection of concurrently running processes. Each process has its own region of memory for code and data and input and output ports. The processes are complete C programs linked together with the runtime library. Within each process there can be one or more concurrently running 'threads'. In Parallel C many instances of the same function can exist simultaneously as threads with each thread having a copy of its own data. Threads can also share global data, channels and other global resources. The way various processes can be distributed over an array of transputers is described in the Section 2.4.4.

---

[3]Root processor signals to all the pipe processors if no more iteration is required

2.4.1   Creating Threads

Threads are processes created by the programs. There can be more than one thread in one processor. Threads can run in parallel or in series. This is controlled by the priority level. Two 'urgent' threads from the same processor will run in series because each one of them is of equally high priority which means that any one of them cannot be interrupted once it starts running. Whichever thread starts first will run to its completion before the second threads starts. On the other hand, if two 'non-urgent' threads are started in the same processor, these two threads can be interrupted at appropriate times and thus some form of pseudo-parallelism is possible between them. If each of the two threads is residing in different processors, then it is possible for them to run in parallel.

Threads are started with the command

$$thread\_start(fn, ws, wssize, flags, nargs, arg1, ..., argn)$$

In this command, fn is the name of the process that is to be run. It can be a computing task as well as a communication task. ws is a pointer to the workspace and $wssize$ is the size of this workspace. The workspace is usually small. The $flags$ parameter specifies whether the thread that is to be started is of 'urgent' or 'non-urgent' priority. $nargs$ specifies the number of arguments that are to be passed to the calling function and $arg1...argn$ are the arguments involved.

Parallel C also contains in its library other command procedures that deal with threads. The $thread\_start$ command described above is the basic command that starts a thread. However, it is also possible to create a thread using the *thread_create* command. This function allows a thread to be started at the same priority as the current thread. It also allows its workspace to be freed after it has been used so that other threads or procedures can takeover this memory space.

Other than thread's creation, Parallel C also has provision to allow the user to stop a thread, restart a stopped thread, or temporarily deschedule a thread. By allowing the user to stop or deschedule a thread, other threads having the same priority as the stopped or descheduled thread are given the opportunity to perform some background computations. In effect, it is possible to create threads having a priority lower than 'non-urgent'. A summary of all the threads commands found in Parallel C is shown in Table (2.2).

Table 2.2: Threads Procedures present in Parallel C.

Procedure Name	Function
thread_create	Creates a simple thread at the same priority of current thread
thread_start	Starts a new thread
thread_deschedule	Deschedules a thread in order to prevent "hogging" of processor
thread_stop	Stops a running thread
thread_restart	Restart a stopped thread
thread_priority	Returns the current thread priority

2.4.2   Interprocess Communication

There will come a time when two or more processes or threads have to communicate with each other. Communication allows data to be transferred from one process to another. For Parallel C, this form of communication is performed using channels. Two tasks can only communicate when there is at least one channel linking between them. Tasks placed on the same processors can have any number of interconnecting channels. Tasks placed on different processors can only be connected where physical wires connect the processor's links. The number of interconnections between tasks on different processors is therefore limited by the number of hardware links each one has. For the T805 transputer model, there are a total of four physical links. Hence, each processor can only be connected to four other processors.

Parallel C allows the user to transfer a word of information (4 bytes) at a time between two tasks or allows the user to transfer a string of bytes (message) between two tasks. Before any data transfer takes place, the communication channel has to be initialized. This indicates that no thread is attempting to to communicate through this channel. This is performed with the *chan_init* command. Later on, if there is one task that needs to send a word (or message) through a link, there must be a receiving task at the receiving end to receive the word (or message). The pairs of commands that do this communication are *chan_out_word* and *chan_in_word* as well as *chan_out_message* and *chan_in_message*. Each of this link has an address associated with it so that the correct path is taken during the data transfer. The addresses of channels associated with the vectors of input and output processes of a Parallel C program are passed to the program through the main program parameter line. This constitutes a little time delay which makes up the initialization time.

Parallel C also has another set of channel communication commands that deal with timed communication. This set of timed communication commands restrict the messages to be transferred within a certain period of time. If the time limit is not up and the data transfer is over, then the data transfer is considered successful. If not, the function will terminate and returns zero to the calling thread.

The complete list of channel communication commands used in the Parallel C library is shown is Table (2.3).

2.4.3   Semaphores

It is likely that there will come a time when one computation task is ready to perform its next set of computations but because of the synchronization involved, it may have to wait for another task before it can continue. In such a case, it will call for some sort of synchronization to prevent conflicts. Parallel C resolves these access conflicts with the concept known as semaphores. A semaphore is a global memory location that each thread may examine and alter in a single uninterruptible instruction. Each semaphore may have a queue of threads waiting on it. When the currently active thread has ceased to need the resource, it will free the resource by signaling the semaphore. This will enable one of the waiting thread to take over the free resource. Upon taking this resource,

Table 2.3: Channel Communication Procedures used in Parallel C.

Procedure Name	Function
chan_init	Initialize a channel
chan_reset	Resets a channel
chan_in_byte	Receive a byte of data from a channel
chan_in_word	Receive 4 bytes (a word) of data from a channel
chan_in_message	Received a chain of bytes from a channel
chan_out_byte	Sends out a byte of data from a channel
chan_out_word	Sends out 4 bytes (a word) of data from a channel
chan_out_message	Sends out a chain of bytes from a channel
chan_in_byte_t	Receive a byte of data from a channel within a certain time period
chan_in_word_t	Receive 4 bytes (a word) of data from a channel within a certain time period
chan_in_message_t	Received a chain of words from a channel within a certain time period
chan_out_byte_t	Sends out a byte of data from a channel within a certain time period
chan_out_word_t	Sends out 4 bytes (a word) of data from a channel within a certain time period
chan_out_message_t	Sends out a chain of words from a channel within a certain time period

this current thread will lock its resource so that other threads are unable to access the resource.

In Parallel C, semaphores are implemented based on the following three commands. *sema_init*, *sema_wait* and *sema_signal*. Before a semaphore can be used, it has to be initialized with the *sema_init* command. If the semaphore is initialized to 0, it means that there is no free resources available and the calling thread has to queue. Once the current task has completed processing and is ready to release its resource for other threads, it will issue a *sema_signal* command. This will increment the value of the semaphore by 1. The thread which is at the bottom of the stack will examine the value of this semaphore, as long as it is not zero, it indicates that free resources are available and this waiting thread can use these resources. Once, it has taken this resource, it will perform a *sema_wait* operation that will decrement the value of the semaphore by 1. If the value of the semaphore is not zero after decrement, it means that there are still further free resources available. This will allow other waiting threads to use these free resources. If not, the semaphore value will remain at zero until a free resource is available.

There is also another set of semaphores commands found in Parallel C that allows the semaphore commands to be called n number of times. The complete set of commands related to semaphores are found in Table (2.4).

2.4.4   Configuration

A Parallel C application is made up of a number of tasks, distributed over a number of processors, and connected together by unidirectional channels. The channels that connect processes on

Table 2.4: Semaphore Related Procedures present in Parallel C.

Procedure Name	Function
sema_init	Initialize a semaphore
sema_signal	Performs the signal operation on a semaphore
sema_wait	Performs the wait operation on a semaphore
sema_signal_n	Repeats the *sema_signal* command n times
sema_wait_n	Repeats the *sema_wait* command n times
sema_test_wait	Tests whether waiting on a semaphore would block an execution

different processors must be associated with physical wires that can carry one channel in each direction. A host processor must be declared that acts as the server. All these processors can be interconnected together using different topologies. The Ring topology, Torus as well as the Hypercubes topologies are some popular topologies adopted by transputers users. In order for the 3L compiler to know exactly what is the topology of the processor network, it is important to create a configuration file. A set of configuration instructions are provided by the 3L compiler. These instructions are used as input to the 3L configuration program which outputs a transputer network program containing a distributing loader. This distribution loader then bootstraps each processor in the network and then loads on it the appropriate C language complied codes.

A sample program in 3L Parallel C is given in Appendix B. In this appendix, a sample configuration file needed by the sample program is also included.

2.5 Performance Metrics and Benchmark Problems

An important issue in parallel implementation is the evaluation of performance. This requires selecting meaningful metrics and benchmark problems. The performance metric chosen in this book is the time for a training epoch. This is because the main thrust of the book is to develop a theoretical model to predict the time for a training epoch and then use the model to find out the optimal mapping for a neural network so as to minimize the time for an epoch. Further the time for one epoch for a given neural network and training set depends only on the mapping method and not on what the problem is or how many epochs it would take for the neural network to converge for that problem. The number of epochs required for overall convergence of BP will be the same whether the algorithm is running on a serial or parallel machine and depends only on the parameters of the algorithm such as the learning rate, momentum rate etc. What a parallel mapping can do, is to reduce the time to complete each epoch but not the total number of epochs required for convergence. So the performance of a parallel implementation can be assessed from the time for one epoch as compared to a serial implementation. This can be expressed as:

$$\frac{T_{epoch}^{serial} - T_{epoch}^{parallel}}{T_{epoch}^{serial}} \times 100$$

where T_{epoch}^{serial} and $T_{epoch}^{parallel}$ are the time for one epoch from the serial and parallel implementations respectively.

In the literature on parallel implementations connection updates per second (CUPS) and *speedup* are also often used as a performance measure. These are related to the time for an epoch and can be calculated from it.

$$CUPS = \frac{(number\ of\ weights\ ) \times (number\ of\ training\ patterns)}{time\ for\ an\ epoch}$$

$$Speedup = \frac{T_{epoch}^{serial}}{T_{epoch}^{parallel}}$$

The benchmark problems must be so chosen as to provide a good insight into the strengths and weakness of the mapping algorithm being evaluated. With this in mind we chose the Encoder [48], NETTALK [49] and Sonar [50] as our benchmark problems and they are described below.

2.5.1 Encoder

The encoder/decoder problem, often known simply as encoder problem is well known in the neural network field. The encoder network consists of an input layer, a hidden layer and an output layer. The structure is often written as $N - X - N$ where N is the number of input and output neurons and X is the number of hidden neurons. There is a connection from every input unit to every hidden unit and from every hidden unit to every output unit. There are no direct connections from input units to to output units. X is normally smaller than N so as to require some encoding or compression effect. The input and out patterns are binary in nature.

The network is usually presented with N distinct input patterns each of which has one input unit set to 1 and all other input units set to zero. The network has to duplicate the input patterns at the output. Since the number of hidden units is smaller than N the network must perform some encoding and decoding operation to produce the correct output. The encoder network is quite flexible in that it allows variation of the network size and the training set size very easily. For example, varying the encoder size, N, allows variation of the network and training set sizes together; varying X while keeping N fixed allows variation of the network size without varying the training set. Fixing N and X and varying the number of patterns presented allows variation of training set size without varying the network size.

Finally if the number of hidden neurons is such that $X = \log_2 N$ then the encoder is called a tight encoder as in this case every input pattern must correspond to one of the N possible 'states' of the hidden layer if each hidden unit can take only a 0 or a 1. Hidden units usually can take many more than just two values so it is even possible to train encoder networks with hidden neurons fewer than $\log_2 N$. The encoder problem is extensively used in the book right from model validation to optimal pattern allocation analysis. The number of hidden units for the encoder networks used in this book is $\log_2 N$. Fig. (2.3) shows the diagram of a typical N-X-N encoder network.

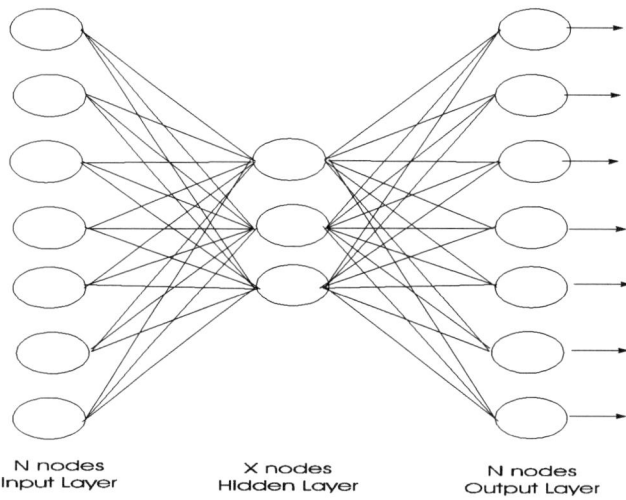

Figure 2.3: A Typical N-X-N encoder network.

2.5.2 NETTALK

The NETTALK is a text to phoneme transcription network used by Sejnowski and Rosenberg [49] in their influential study of speech generation using a neural network. Their data set consisted of 20,008 English words along with a phonetic transcription for each word. The task for the neural network is to learn to produce the proper phonemes, given a string of letters as input.

The input to the network is made up of a series of seven consecutive letters from one of the training words. For each of the seven letter positions in the input the network has a set of 29 input units: one for each of the 26 letters in English and three for punctuation characters. Thus there are $29 \times 7 = 203$ input units in all.

The output side of the network uses a distributed representation for the phonemes. There are 21 output units representing various articulatory features such as voicing and vowel height. Each phoneme is represented by a distinct binary vector over this set of 21 units. In addition there are 5 output units that encode stress and syllable boundaries.

Sejnowski and Rosenberg did not use a fixed number of hidden units. They varied these from one experiment to another from 0 to 120. The number of hidden units for NETTALK networks used in this book is 60. Fig. (2.4) shows the NETTALK network that is used in this study.

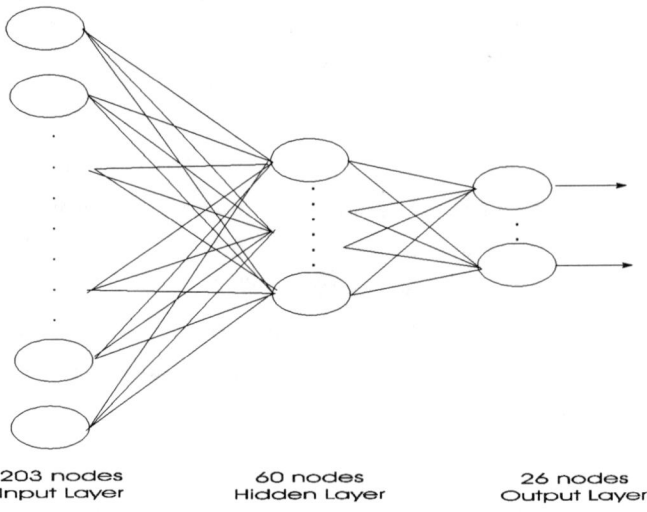

203 nodes   60 nodes   26 nodes
Input Layer   Hidden Layer   Output Layer

Figure 2.4: The NETTALK network.

2.5.3   Sonar

The Sonar problem is a study on classification of sonar signals using neural networks by Gorman and Sejnowski [50]. The task for the network is to learn to discriminate between sonar signals bounced off a metal cylinder and those bounced off from a roughly cylindrical rock.

The training set used by Gorman and Sejnowski had 208 patterns. 111 patterns were obtained by bouncing sonar signals off a metal cylinder at various angles under various conditions. The remaining 97 patterns were obtained from rocks under similar conditions. The transmitted sonar signal is a frequency-modulated chirp rising in amplitude. The data set contained signals obtained from a variety of aspect angles spanning 90 degrees for the cylinder and 180 degrees for the rock.

The neural network for the Sonar problem has 60 input units and two output units one to indicate the metal cylinder and the other to indicate a rock. Gorman and Sejnowski did not use a fixed number of hidden units. These were varied for different experiments from 2 to 24. The Sonar benchmark problem used in the book has 60 input units, 24 hidden units and 2 output units. The number of patterns in the training set is 192. A diagram of the Sonar network used in this book is shown in Fig. (2.5).

The NETTALK and Sonar problems are used in Chapters 5 and 6 to assess the performance of optimal pattern allocations.

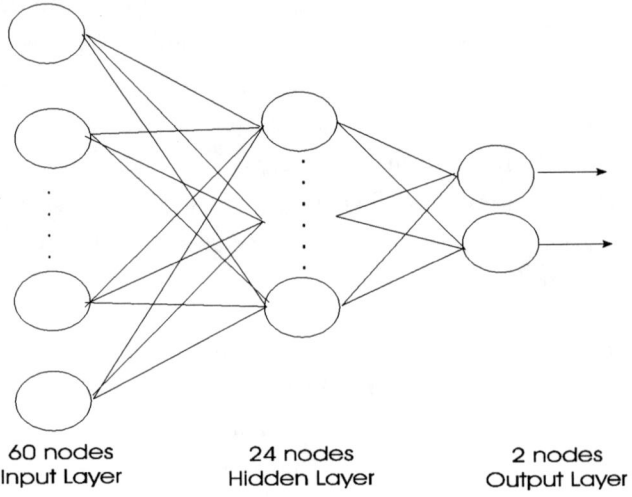

Figure 2.5: The Sonar network.

Chapter 3

Development of a Theoretical Model for Training Set Parallelism in a Homogeneous Array of Transputers

Given a certain neural network configuration and a number of training samples, it will be nice if we can predict the time needed for one training epoch of the backpropagation algorithm. This will give us a rough idea of the expected improvement obtained from parallelism. In this chapter, we develop a theoretical model that predicts the time needed to compute an epoch [51]. Pipeline-ring topology for the transputer array and Training Set parallelism are assumed in the formulations.

To perform this analysis, the time taken for each of the processes in the backpropagation learning algorithm is worked out. The processes and the time taken for them consist of the time taken to accumulate the weight changes, the time taken for the forward pass and backward pass, as well as the time to transfer the weight changes, etc. These timings are worked out by breaking these processes to simple mathematical operations and knowing the timings for these elemental operations which can be measured directly, the training time per epoch can be worked out. These elemental operations consist of 'Add and Assign', 'Multiply and Assign', etc. and other basic communication operations such as the time taken

37

to initialize a communication process and the time taken to send or receive a certain block of data.

This chapter first covers the basic timing components of a transputer implementation. Here we will discuss the main time components that make up any parallel implementation on transputers. The main time components include the time needed for communications between transputer links and the time needed for computations and these are discussed in Section 3.1. Section 3.2 outlines the main steps needed to obtain the time per epoch for the backpropagation algorithm. Section 3.3 covers the time components for the parallelized Backpropagation Algorithm. This section is broken down into two subsections. They are Section 3.3.1 and Section 3.3.2. In Section 3.3.1, the time components related to the data communication between the links of the transputers is discussed. Section 3.3.2 deals with the computation part of the backpropagation algorithm. In this section, the various time components for the forward pass and the backward pass of the backpropagation algorithm is discussed. Section 3.4 summarizes the experimental results for validating the theoretical T_{epoch} model.

3.1   Time Components of Parallel Transputer Implementation

In this section, the time components for a parallel implementation of transputers are discussed. Broadly, there are two time components present, the first being the communication time component and the second being the computation time component. Communication time is the time taken to transfer blocks of data over the transputer links whereas computation time is the time taken to calculate the operations that make up the process.

In the following analysis, it should be emphasized that the equations developed below are also applicable to any pipeline processor parallel system. In this analysis, we assume that the whole neural network can be placed into each of the transputers connected in a pipeline-ring topology as shown earlier in Fig. (2.2). In this implementation, in order to minimize communication time, the weights are not updated after every training pattern. Rather, they are updated only after the whole set of training patterns has been shown to the network. This form of updating is known as epoch update. Hence for our timing equations, we will be measuring the time taken for the network to calculate one epoch.

In any transputer implementation, there are two main time components that constitute the total time taken per iteration, T_{iter}. One component is known as the communication time, T_{comm} and the other is known as the computation time, T_{comp}.

That is,

$$T_{iter} = T_{comm} + T_{comp} \tag{3.1}$$

T_{comm} includes the time taken to initiate all the communication processes as well as the total time taken to transfer (send or receive) data across the transputers' links. Therefore,

$$T_{comm} = T_k + T_v \qquad (3.2)$$

where T_k is the time taken to initiate all the communication processes and T_v is the time taken to send or receive packets of data across the transputer network.

The time spent on computation T_{comp} also has a constant part T_{con} and a variable part T_{var}. T_{con} is the time spent on such things like initializing the variables (e.g. weight changes) at the start of an iteration and T_{var} is the time taken to perform the training calculations. Increasing the number of processors in a network, for instance, would lead to a decrease in T_{var} but not in T_{con} .

Numerically,

$$T_{comp} = T_{con} + T_{var} \qquad (3.3)$$

Taking the two timing components together, the time taken for one iteration can be expressed in the following form.

$$T_{iter} = T_k + T_v + T_{con} + T_{var} \qquad (3.4)$$

For the problem discussed in this study, we are required to find out the time per epoch for the backpropagation algorithm (T_{epoch}). With respect to the total time taken per iteration, T_{iter} for the transputers, we have

$$T_{iter} = T_{epoch} \qquad (3.5)$$

By finding out the actual time taken for each of the components in Eq. (3.4), we can find the time taken to calculate one epoch of the backpropagation algorithm.

3.2 Timing Aspects of Parallelizing the Backpropagation Algorithm

In this section, an outline of the timing aspects of parallelizing the backpropagation algorithm is discussed. In other words, we are relating the time taken per epoch of the backpropagation algorithm with the generic time equation, Eq. (3.4) in Section 3.1.

The steps for calculating one epoch of the backpropagation algorithm in each transputer as well as the symbols used to describe the time taken for each operation are summarized below:

1. The weight changes and bias changes for the current epoch are initialized to zero at the start. The time taken for this operation is defined as $T_{con}^{\Delta w_0}$.

2. After initialization, the forward pass and the backward pass of the backpropagation algorithm is performed for each training pattern assigned to the processor. The forward pass involves calculation of the product sums of weights and activation and sigmoid operations at each neuron in the network. The quadratic error between the desired and actual output is also calculated. The time taken for this operation is defined as T_f. In the backward pass the weight change due to each

pattern is calculated. The time needed for this operation is T_b.

3. The total weight change and error due to all the patterns assigned to the transputer are calculated by accumulating the weight change and error calculated for individual patterns in step 2. It takes $T_{con}^{\Delta w}$ amount of time for this operation to complete in each processor.

4. The first pipe processor sends its accumulated weight change and error to its downstream (right) neighbor which adds its own accumulated weight change and errors and sends the updated sums to its right neighbor which in turn updates the sums and sends to its right neighbor and so on. The last pipe processor would thus receive the accumulations of all the processors to its left, adds its own weight change and error and obtains the total weight change and error for the epoch. It then adds the weight change to the existing weight values and obtains new values for all weights in the neural network. The new weight values are then passed upstream by each pipe processor except the first. In an n transputer network, there will be $n - 1$ transfers of weight change and error and another $n - 1$ transfers of weights. The time taken to send or receive one set of weight change and error from a upstream pipe processor to a downstream pipe processor is $T_{init} + T_{trf}^{\Delta w}$. The time taken to send or receive one set of updated weights back from a downstream pipe processor to a upstream pipe processor is $T_{init} + T_{trf}^{w}$. T_{init} refers to the time taken to initialize or start off a communication process while $T_{trf}^{\Delta w}$ and T_{trf}^{w} represent the time taken to transfer a packet of weight changes or weights across two communication channels

respectively.

5. At the same time as the weights are updated, the last processor sends the total error to the root processor which checks whether convergence is reached. The time taken for this operation is small and is neglected.

In our implementation of the backpropagation algorithm all the processors are connected in a pipeline-ring architecture. In a pipeline-ring architecture, each processor is connected to an upstream and a downstream neighbor. Messages are transferred from one processor to another.

The timing diagram for an epoch is as shown in Fig. (3.1). In this figure, $P_1, \ldots, P_n$ are the number of patterns allocated to the processors. $1, \ldots, n$. In this figure, the arrows show how the communication tasks are being synchronized with their neighboring processors. It is observed that the processors are connected in a pipelined-ring topology. This is because the communications only occurs between two successive processors. While the weight changes and errors are being accumulated, the transfers are from left to right. Once the nth processor has received and accumulated the weight changes and errors from the $(n-1)^{th}$ processor, the transfers change from right to left as indicated by the arrowheads.

The various notations used are explained below:

- $T_{con}^{\Delta w_0}$: Time for initializing the weight changes and error.

- T_c : Time taken to perform the forward and backward pass of backpropagation for a single pattern.

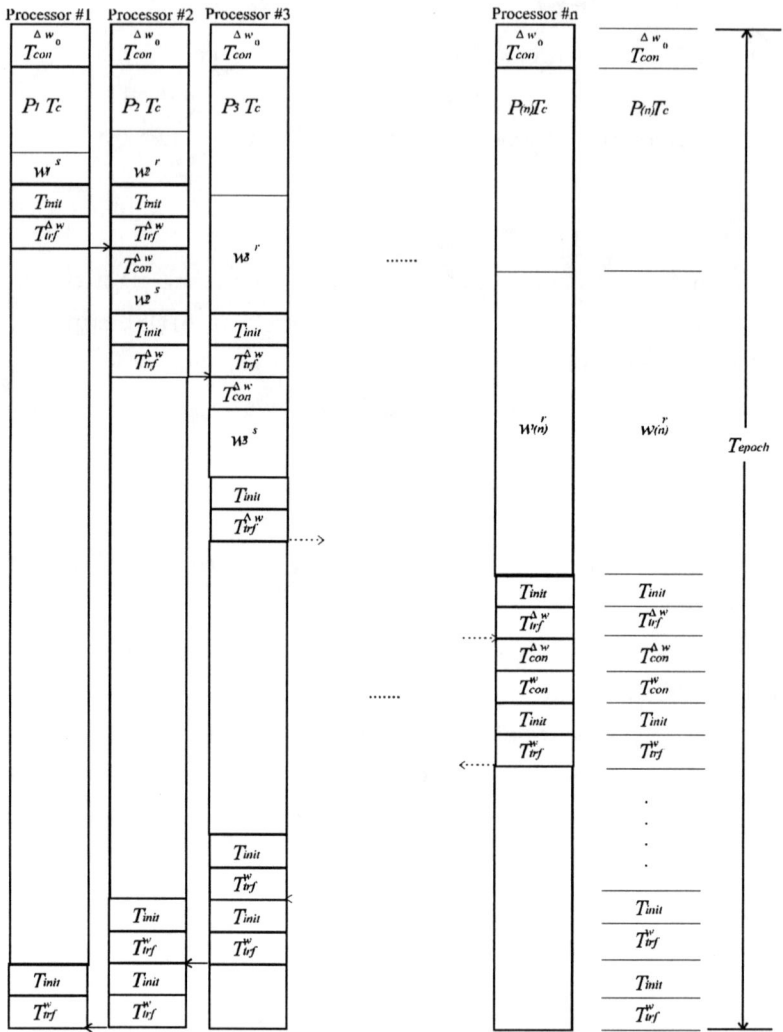

Figure 3.1: Timing Diagram for an epoch for a homogeneous array of transputers.

- $w_{(j)}^s$ is the waiting time required before the weight changes can be sent to the next downstream processor. This delay occurs when the sending task is ready to send but the receiving task is still performing some calculations causing the sending task to wait.

- $w_{(j)}^r$ is the waiting time incurred in the receiving task before the weight changes can be received. This situation occurs when the receiving task is ready to receive data from the sending task, but the sending task is still busy with the calculations causing asynchronization in the two tasks. It can be observed that although there is always a pair of 'waiting' timings in each communication process, however, at any one time, one value of the 'waiting' pair is always zero. Another observation is that unlike processors 2 to $n - 1$, there is only one waiting time in the first and last processors. This is because there is no need to receive any weight changes for the first processor and similarly, no weight changes are needed to be sent down the stream by the last processor.

- T_{init} : Time taken to start off a communication process.

- $T_{trf}^{\Delta w}$: Time spent for sending the weight changes and errors.

- $T_{con}^{\Delta w}$: Time needed to accumulate the weight changes and errors.

- T_{con}^{w} : Time spent on updating the weight values in the last pipe processor.

- T_{trf}^{w} : Time spent on sending the updated weights.

From the timing diagram in Fig. (3.1), it can be seen that the overall time per epoch depends on the total time taken by the nth processor (last processor in the pipeline).

i.e. The time per epoch is

$$T_{epoch} = T_{con}^{\Delta w_0} + P_{(n)}T_c + w_n^r + T_{init} + T_{trf}^{\Delta w} + T_{con}^{\Delta w} + T_{con}^{w} + (n-1)[T_{init} + T_{trf}^w] \tag{3.6}$$

Since the time taken to transfer a set of weight changes and error is approximately the same as the time taken to transfer a set of weights, we can simplify Eq. (3.6) to

$$T_{epoch} = T_{con}^{\Delta w_0} + P_{(n)}T_c + w_n^r + T_{con}^{\Delta w} + T_{con}^{w} + n[T_{init} + T_{trf}] \tag{3.7}$$

where T_{trf} is the time taken to transfer a set of weight changes or weights.

3.3   Time Components for the Parallelized Backpropagation Algorithm

Calculation of the various components of the time per epoch (T_{epoch}) is given in this section. Rearranging T_{epoch} as given in Eq. (3.7) we see that it consists of a communication part and a computation part as shown below.

$$T_{epoch} = \underbrace{n[T_{init} + T_{trf}]}_{T_{comm}} + \underbrace{T_{con}^{\Delta w_0} + P_{(n)}T_c + w_n^r + T_{con}^{\Delta w} + T_{con}^w}_{T_{comp}} \quad (3.8)$$

3.3.1 Communication part [T_{comm}]

Based on Eq. (3.2) found in Section 3.1, it can be seen that if we relate this equation with Eq. (3.8), we have $T_k = nT_{init}$ and $T_v = nT_{trf}$. T_{init} is the time required to initiate a communication process and T_{trf} is the time taken for the transfer of data; i.e. data such as weight and bias changes as well as updated weights and biases. Since number of weight and bias changes equals number of weights and biases,

$$T_{trf} = rdN_{wb}$$

where N_{wb} is the total number of weights and biases in the neural network. r is the precision of the representation of a single data in bytes ($r = 4$ for a T805 transputer), d is the time taken to send or receive one byte over a link. For our transputer array, rd is equal to 2.25μ sec. Since the number of biases is the same as the number of neurons in the network $N_{wb} = N_w + N_{tot}$, where N_w is the total number of weights in the neural network and N_{tot} is the total number of neurons (excluding input nodes) in the network. The communication part (T_{comm}) can now be written as

$$T_{comm} = n[T_{init} + rdN_{wb}] \quad (3.9)$$

3.3.2   Computation part $[T_{comp}]$

Referring to Eq. (3.3) in Section 3.1 and Eq. (3.8), we can see that the computation part itself (T_{comp}) can be further divided into two parts. A constant part (T_{con}) which does not change with respect to the number of processors in the topology and a variable part (T_{var}) that varies with respect to the number of processors in the topology.

i.e. $T_{comp} = T_{con} + T_{var}$

The time taken for the constant and variable part are calculated in terms of basic mathematical operations. The basic operations such as the time to accumulate two two-dimensioned floating numbers together, time to multiply two two-dimensional floating numbers together and other basic mathematical operations are measured experimentally and are given in Table (3.1). Since these timings will be used often, it is worthwhile to note that the symbols for denoting these timings are defined as $t_{abbreviated\ elemental\ operation}$. For example, the elemental time for addition is t_a.

(A) Constant part (T_{con})

From Eq. (3.8), it can be seen that

$$T_{con} = T_{con}^{\Delta w_0} + T_{con}^{w} + T_{con}^{\Delta w} \qquad (3.10)$$

Table 3.1: Time taken to perform various mathematical operations for transputers and the 486-33 processor.

Symbol	Description	Operation in C language	Time taken ($\mu secs$)		
			T805-20	T805-25	486-33
t_m	Multiply two different floats	A*B	1.565	1.252	0.934
t_{ma}	Multiply and assign two different floats	A=B*C	6.363	5.067	1.319
t_{mas}	Compound multiplication two same floats	A*=B	6.171	4.950	1.264
t_a	Add two different floats	A+B	1.565	1.252	0.879
t_{aa}	Add and assign two different floats	A=B+C	6.363	5.067	1.209
t_{aas}	Compound addition of floats	A+=B	6.171	4.950	1.319
t_s	Subtract two different floats	A-B	1.416	1.129	0.934
t_{sa}	Subtract and assign two different floats	A=B-C	6.434	5.143	2.637
t_{sas}	Compound subtraction two same floats	A-=B	6.240	4.985	1.978
$t_{sigmoida}$	Sigmoid calculation with assignment	A= SIGMOID(B)	81.969	65.520	29.231
t_{grad}	Derivative of sigmoid	GRAD(A)	8.229	6.583	2.692
t_{ass}	Assignment	A=B	5.143	4.114	0.879
T_{init}	Initiate a communication process	N.A.	9.00	9.00	N.A.

The time taken to compute each component of T_{con} is described below.

(i) Time taken to initialize the weight and bias changes
$(T_{con}^{\Delta w_0})$

At the start of an epoch all the weight changes and bias changes
have to be initialized to zero to nullify the values existing from the
previous epoch. The initialization requires one assignment state-
ment to perform the assign zeros to each of the weight changes and
bias changes. Since there are N_w weights and N_{tot} biases in the
network, $T_{con}^{\Delta w_0}$ is given by :

$$T_{con}^{\Delta w_0} = (N_w + N_{tot})t_{ass} \qquad (3.11)$$

where t_{ass} represents the elemental operation time for assignment and is given in Table (3.1).

**(ii) Time taken for updating the weight values in the
last pipe processor** (T_{con}^{w})

This weight updating is performed only in the last pipe processor. The 'C' like codes for performing the weight update are as
follows.

```
for i = 1; i < Nw + 1; i++  {
/* update the weights */
wj+1[i]  =  β * Δwj[i] + Δwj+1[i] + wj[i]
/* Since momentum term is used we need */
/* the weights from the previous epoch.   */
```

```
/* So an assignment statement is required */
/* to store the weights of current epoch */
/* for use in the next round.  */
```

$w_j[i] = w_{j+1}[i]$

```
}
for i = 1; i < N_tot + 1; i++ {
/* accumulate the biases */
```

$Bias_{j+1}[i] = \beta * \Delta Bias_j[i] + \Delta Bias_{j+1}[i] + Bias_j[i]$

$Bias_j[i] = Bias_{j+1}[i]$

```
}
```

In the above $j+1, j$ refers to the current and previous epochs and β refers to the momentum term present in the backpropagation algorithm. The time taken for updating the weights and biases are:

$$T_{con}^w = (N_w + N_{tot})(t_m + t_a + t_{aa} + t_{ass}) \qquad (3.12)$$

where t_m, t_a and t_{aa} represent timings for elemental operations as defined in Table (3.1).

(iii) Time taken for accumulating the weight changes, bias changes and errors $(T_{con}^{\Delta w})$

Each pipe processor has to accumulate its own weight and bias changes to the weight and bias changes it has received from its upstream neighbor before sending them to its downstream neighbor.

The 'C' codes for the accumulation of weight and bias changes and the error are as follows:

```
for i = 1; i < N_w + 1; i++  {
/* accumulate weight changes */
Δ w [i] + = Δ w [i]
}
for i = 1; i < N_tot + 1; i++  {
/* accumulate bias changes */
Δ Bias [i] + = Δ Bias [i]
}
/* accumulate errors */
error += error
```

The time taken to accumulate weight and bias changes and the error is given by :

$$T_{con}^{\Delta w} = (N_w + N_{tot})t_{aas} + t_{aas} \qquad (3.13)$$

The time for the constant part of computation can now be written as,

$$T_{con} = N_{wb}[3t_{ass} + t_m + t_a + t_{aa}] + t_{aas} \qquad (3.14)$$

(B) Variable part (T_{var})

From Eq. (3.8), it can be seen that the variable part of computation is

$$T_{var} = P_{(n)}T_c + w_n^r \qquad (3.15)$$

where T_c is the time taken for one forward T_f and backward T_b pass of the backpropagation algorithm.

i.e. $T_c = T_f + T_b$.

w_n^r is the waiting time found in the last processor. This waiting time w_n^r can only be determined after the allocation of the patterns in each processors is known. More details of w_n^r will be discussed in Section 4.1 of Chapter 4.

(i) Time taken for the forward pass (T_f)

The forward pass consists of the following parts:

1. Time taken to assign input values and desired output values to the input and output units of the neural network, T_{f1}.

   This part is made up of assignment statements as shown by the code below.

```
for i = 1; i < N₁ + 1; i++ {
/* assign input values of the current */
/* pattern */
/* to input units */
a[i] = input[i]
}
for i = 1; i < N₁ + 1; i++ {
```

```
/* assign desired output values for the
/* current pattern to the output units */
o[i] = desired-output[i]
}
```

N_1 is the number of units in the first (i.e. input) layer and N_l is the number of units in the last (i.e. output) layer. The time taken for T_{f1} is

$$T_{f1} = (N_1 + N_l)t_{ass}$$

2. Time to calculate the total input to all the units in all layers (except input layer), T_{f2}.

The total input to unit i in the $k + 1$ layer $(net_{k+1}[i])$ is the sum of the product of weight and activation at each fan-in of unit i as given below.

$$net_{k+1}[i] = \sum_{j=1}^{N_k} a_k[j]w_k[i][j] \quad + b_k[i]$$

where $a_k[j]$ is the activation value of the unit j in the layer k, $w_k[i][j]$ is the weight connecting unit j in the layer k to the unit i in layer $k + 1$, $b_k[i]$ is the bias to the unit i and N_k is the total number of units in the layer k. The following code implements the above equation for all nodes in $k + 1$ layer.

```
for i = 1; i < N_{k+1} + 1; i + +  {
/* initialize net_{k+1}[i] to zero */
net_{k+1}[i] = 0
/* calculate the sum of products for */
/* each unit */
```

```
/* in layer k + 1 */
for i = 1; i < Nk + 1; i ++  {
netk+1[i] += ak[j] * wk[i][j]
}
/*Add bias term */
netk+1[i] += bk[i]
}
```

It can be seen from the code that the time taken to calculate the total input to a single unit i in the $k+1$ layer is $N_k(t_m + t_{aas}) + t_{aas} + t_{ass}$.

The time taken to calculate the total inputs for all the units in the $k+1$ layer is T_{k+1} given by the following equation.

$$T_{k+1} = N_{k+1}[N_k(t_m + t_{aas}) + t_{aas} + t_{ass}]$$

The time for calculating T_{f2} is $\sum_{k=1}^{l-1} T_{k+1}$. Substituting for T_{k+1} and with some simplification, T_{f2} can be written as

$$T_{f2} = (t_m + t_{aas}) N_w + (t_{aas} + t_{ass})N_{tot}$$

3. Time to calculate the activation level at every unit in all the layers except input layer, T_{f3}.

The activation function used is the standard sigmoid and time T_{f3} is given by

$$T_{f3} = N_{tot}\, t_{sigmoida}$$

4. Time to calculate the error at the output units, T_{f4}.

The error at the output units is given by

$$error = \sum_{i=1}^{N_l}(a_l[i] - o_l[i])(a_l[i] - o_l[i])$$

where a_l, o_l are the actual and desired output values. The code below implements the above equation.

```
for i = 1; i < N_l + 1; i++  {
/* calculate the output error */
error + = ( a_l[i] - o_l[i] ) * ( a_l[i] - o_l[i]
)
}
```

The time taken for calculation of the error is given by

$$T_{f4} = N_l(t_s + t_m + t_s + t_{aas})$$

Summing up all the components, the time taken for a forward pass is :

$$T_f = T_{f1} + T_{f2} + T_{f3} + T_{f4} \qquad (3.16)$$

(ii) Time taken for the backward pass (T_b)

The backward pass consists of the following parts:

1. Time to calculate the deltas at the output units, T_{b1}

   For the unit i in the output layer, $\delta_l[i]$ is

$$\delta_l[i] = (a_l[i] - o_l[i]) \, f'[net_l(i)]$$

where $f'[net_l(i)]$ is the derivative of the sigmoidal function. The following code shows the implementation of the calculation of deltas for all units of the output layer l.

```
for i = 1; i < N_l + 1; i + + {
/* calculate the deltas at the output nodes
*/
δ_l[i] = (a_l[i] − o_l[i]) * f'[net_l(i)]
}
```

The time taken to calculate the deltas at the output nodes is

$$T_{b1} = N_l \left(t_s + t_{grad} + t_{ma} \right)$$

2. Calculation of deltas at the hidden units, T_{b2}.

For the unit i in the hidden layer k, $\delta_k[i]$ is

$$\delta_k[i] = \left(\sum_{j=1}^{N_{k+1}} \delta_{k+1}[j] \, w_k[j][i] \right) f'[net_k(i)]$$

The code for the above is

```
for i = 1; i < N_k + 1; i + + {
/* initialize δ_k[i] to zero.*/
δ_k[i] = 0
for j = 1; j < N_{k+1} + 1; j + + {
δ_k[i] + = δ_{k+1} [j] *w_k[j][i]
}
δ_k[i] * = f'[net_k(i)]
}
```

The time taken to calculate $\delta_k[i]$ for the unit i is $N_{k+1}(t_m + t_{aas}) + (t_{grad} + t_{mas})$. Time to calculate the $\delta_k[i]$ for all units in the k layer is given by T_k where

$$T_k = N_k[N_{k+1}(t_m + t_{aas}) + (t_{grad} + t_{mas} + t_{ass})]$$

The time taken to calculate the deltas for all hidden units in the network is

$$T_{b2} = \sum_{k=2}^{l-1} N_k[N_{k+1}(\, t_m + t_{aas}\,) + (t_{grad} + t_{mas} + t_{ass})]$$

After simplification T_{b2} can be expressed as

$$T_{b2} = (t_m + t_{aas})(N_w - N_1 N_2) + (t_{grad} + t_{mas} + t_{ass})(N_{tot} - N_l)$$

3. Calculation of weight changes, T_{b3}

   The code to calculate the weight change for a single weight is

   $$\delta w[j][i] + = \alpha * \delta[j] * a[i]$$
   ```
   /* α is the learning rate */
   ```

   So the time to calculate the weight change for all the N_w weights will be,

   $$T_{b3} = N_w(2t_m + t_{aas})$$

4. Calculation of bias changes, T_{b4}

   The code to calculate the change for a single bias is

   $$\Delta Bias[i] + = \alpha \delta[i]$$

   So the time to calculate the change for all the N_{tot} biases will be,

   $$T_{b4} = N_{tot}(t_m + t_{aas})$$

   Summing up all the components listed above for a backward pass

$$T_b = T_{b1} + T_{b2} + T_{b3} + T_{b4} \tag{3.17}$$

Now, the time taken for one forward pass and one backward pass (T_c) can be written as :

$$T_c = \underbrace{(T_{f1} + T_{f2} + T_{f3} + T_{f4})}_{T_f} + \underbrace{(T_{b1} + T_{b2} + T_{b3} + T_{b4})}_{T_b}.$$

Substituting for $T_{f1}, T_{f2}, T_{f3}, T_{f4}, T_{b1}, T_{b2}, T_{b3}$ and T_{b4} and after simplification,

$$
\begin{aligned}
T_c \;=\; & (4t_m + 3t_{aas})N_w + t_{ass}N_1 - (t_m + t_{aas})N_1N_2 \\
& + (2t_{aas} + t_m + t_{sigmoida} + t_{grad} + t_{mas} + 2t_{ass})N_{tot} \\
& + (3t_s + t_m + t_{aas} + t_{ma} - t_{mas})N_l. \quad\quad (3.18)
\end{aligned}
$$

Combining Eqs. (3.15), (3.9), (3.14) and (3.15) gives the time for an epoch (T_{epoch}) as

$$
\begin{aligned}
T_{epoch} \;=\; & n[T_{init} + r d N_{wb}] + \\
& P_{(n)}T_c + w_n^r + \\
& N_{wb}[3t_{ass} + t_m + t_a + t_{aa}] + t_{aas} \quad\quad (3.19)
\end{aligned}
$$

If we relate Eq. (3.19) with the more general Eq. (3.4), we have,

$$
\begin{aligned}
\underbrace{T_{epoch}}_{T_{iter}} \;=\; & \underbrace{nT_{init}}_{T_k} + \underbrace{nrdN_{wb}}_{T_v} + \\
& \underbrace{N_{wb}[3t_{ass} + t_m + t_a + t_{aa}] + t_{aas}}_{T_{con}} + \\
& \underbrace{P_{(n)}T_c + w_n^r}_{T_{var}} \quad\quad (3.20)
\end{aligned}
$$

The above equation gives the time per epoch for backpropagation algorithm implemented in a pipelined ring transputer array and is defined based on elemental mathematical operations. It also clearly defines the different communication and computing components in the parallel implementation. It is also useful to recollect the equation for T_{epoch} which was described in an overall way i.e. Eq. (3.7).

$$T_{epoch} = T_{con}^{\Delta w_0} + P_{(n)}T_c + w_n^r + T_{con}^{\Delta w} + T_{con}^w + n[T_{init} + T_{trf}] \quad (3.21)$$

3.4   Validation of the T_{epoch} Model

The above equation gives the T_{epoch} for parallel implementation of backpropagation neural networks implemented in a homogeneous array of transputers in a pipelined ring. It is based on timings of elemental operations in a transputer. Before using this model for other purposes, it is important to carry out a validation test. This can be done by allotting different pattern allocation schemes and using the theoretical model to predict the T_{epoch} and then comparing the same by running in a transputer array. This validation has been carried out for the encoder benchmark problem [48] described in Section 2.5.1 of Chapter 2. The encoder problem is chosen because of its popularity and flexibility as it allows varying the neural network size and training set size very easily.

Our experiment involves mapping a 256-8-256 encoder network with 256 input patterns and 4360 connections including biases on to a homogeneous processor network with 10 T805-20 transputers.

For each of this experiment, each transputer is allocated with $25 \pm \sigma$ number of training patterns where σ is a uniformly distributed integer between -12 and 12. Each of these patterns distribution will then be normalized so that the sum of all patterns is equal to 256. For example, in Table (3.2) which shows the time per epoch in seconds for the encoder problem, experiment 1 has pattern distribution 13, 14, 23, 15, 27, 23, 35, 24, 48 and 34, while experiment 2 has pattern distribution 35, 18, 24, 15, 31, 27, 14, 20, 47 and 25. The experimental time per epoch for these two cases are 5.789 sec and 5.672 sec respectively. Likewise, for all the other 18 cases, each one of them has its own random pattern distributions. In this table, the first column gives the experimental time per epoch (sec), second column gives the theoretical time per epoch (sec), and the third column gives [(theory - expt) / theory] %

The bottom two rows give the average and standard deviation of the percentage errors respectively.

After analyzing the results, we notice that there are some differences between the theoretical and experimental results these differences may be due to the following factors:

- In the theoretical calculations, two dimensional array timings are used in all cases when the basic mathematical timings are being substituted. It can be seen that in the implementation of the backpropagation algorithm, a majority of the calculations involve weights and biases. Two dimensional arrays are usually used to store these weights and biases variables. Thus only these variables will use two dimension array timings in experiments.

Table 3.2: Comparison between Experimental and Theoretical Results.

Encoder			
No.	Experimental(sec)	Theoretical(sec)	Difference (%)
1	5.788786	6.098737	5.082206
2	5.671605	5.976848	5.107088
3	4.201519	4.454057	5.669828
4	4.169595	4.405629	5.357557
5	4.456835	4.721228	5.600094
6	4.646098	4.916578	5.501393
7	4.764664	5.038467	5.434266
8	5.389782	5.684642	5.186963
9	5.871762	6.183895	5.047517
10	4.810545	5.086895	5.432575
11	4.410894	4.6728	5.604917
12	4.810568	5.086895	5.432121
13	4.989813	5.270548	5.326493
14	4.48345	4.746262	5.537238
15	4.643582	4.893185	5.101024
16	4.928525	5.208784	5.380509
17	4.406596	4.649407	5.2224
18	4.827666	5.100232	5.344189
19	4.328623	4.587642	5.646033
20	4.729184	5.001736	5.449151
Average			5.373178
Std. Dev			0.195182

Substituting two dimensional array timings in all the theoretical equations, which we had done, will cause the theoretical results to be higher than the experimental results because in the actual "C" codes, a combination of both single and double dimension arrays are being used.

- The time taken to set the indices of the pointers of the variables are not included in the theoretical model.

- The time taken to check for the satisfaction of certain error criteria is not included in the formulation. These error checkings are performed by the first task (master) when the last downstream task is performing its updating of the current set of weights.

Based on these results, it can be concluded that the model for predicting the T_{epoch} for a pipelined ring of transputers give an estimate which has been verified and found to be within 6% of the experimental timings.

Chapter 4

Equal Distribution of Patterns Amongst a Homogeneous Array of Transputers

In Chapter 3, we have developed a model to predict the time per epoch for a homogeneous array of transputers in term of its elemental timings, specifically Eq. (3.19). Using this equation, one can predict the time per epoch given *any distribution* of training patterns. In this chapter, we develop the time per epoch for a special case where the patterns are distributed equally among all the transputers in the array. Equal distribution of patterns in an array is the most straightforward allocation one can think of when the processors are the same i.e. an homogeneous array. Hence, it is instructive to check whether the developed model is accurate in this case. Also, using the above model for this case we can say more like finding the optimal number of transputers to obtain the minimum time per epoch and the associated cost-benefit analysis. This really helps in assessing whether we can minimize the training time for a particular problem just by going on increasing the number of transputers in the array [52]. Alternatively, this kind of analysis also indicates the number of processors (a bound on the number of transputers) above which the benefits are marginal.

4.1   Analytical Model for Time per Epoch

For the equal distribution case considered here, we set $P_1, \ldots, P_n$ to P_{tot}/n. With this equal allocation, the timing diagram is simplified to that shown in Fig. (4.1).

In this figure, $T_{cal}^p = (P_{tot}/n)T_c$. It is the time taken for the forward and backward pass of the backpropagation for P_{tot}/n assigned patterns in each processor. By comparing this timing diagram with the timing diagram in Fig. (3.1), we notice that we can simplify Eq. (3.19). After simplifying, we notice that

$$P_{(n)}T_c = T_{cal}^p = (P_{tot}/n)T_c \qquad (4.1)$$

and

$$w_n^r = (n-2)T_{init} + (n-2)T_{trf}^{\Delta w} + (n-2)T_{con}^{\Delta w} \qquad (4.2)$$

In other words, the waiting time w_n^r present in the last processors is due to the time taken to transfer and accumulate the weight changes from the first processor to the $(n-1)^{th}$ processor.

If we now substitute Eqs. (4.1) and (4.2) into Eq. (3.19) and simplify, we have, the time per epoch for the equal distribution of patterns among a processor array as:

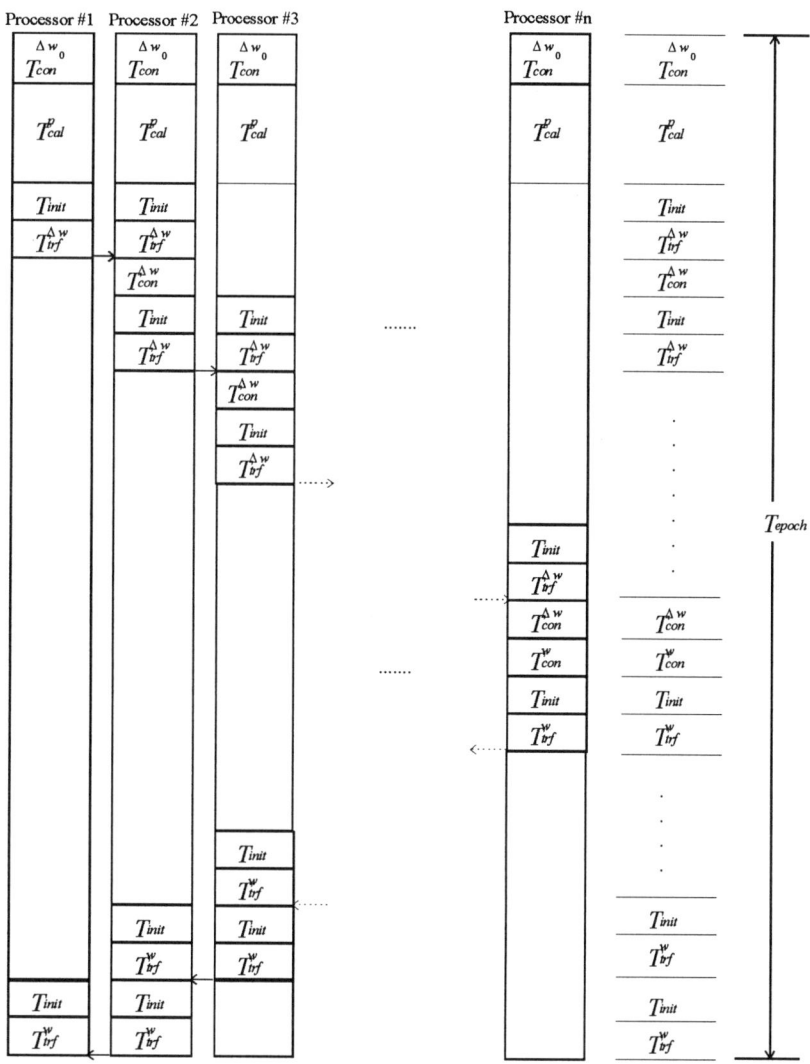

Figure 4.1: Timing Diagram for equal distribution of patterns in a homogeneous array of processors.

$$T_{epoch} = 2(n-1)[T_{init} + rdN_{wb}]$$
$$+ \frac{P_{tot}T_c}{n} + (n-1)t_{aas}$$
$$+ N_{wb}[2t_{ass} + t_m + t_a + t_{aa} + (n-1)t_{aas}] \quad (4.3)$$

4.2   Validation of the Model for Equal Distribution

We have derived the model for predicting the T_{epoch} when the patterns are distributed equally among all the transputers in the pipelined ring array. Before using this model for any meaningful purpose, it is important to carry out validation tests on the model by conducting experiments on a transputer array. The benchmark chosen here is the encoder problem described earlier in Section 2.5.1.

Fig. (4.2) shows the theoretically calculated (from Eq. (4.3)) and experimentally obtained values for T_{epoch} for a 64-6-64 encoder with varying training set sizes. The processor network in this experiment had four T805-20 transputers each with 1 Mbytes of external memory.

Fig. (4.3) shows the theoretical and experimental values for T_{epoch} for a 256-8-256 encoder for varying number of (T805-20) transputers.

As can be seen from Figs. (4.2) and (4.3) the theoretically predicted values for T_{epoch} are close to the experimental results

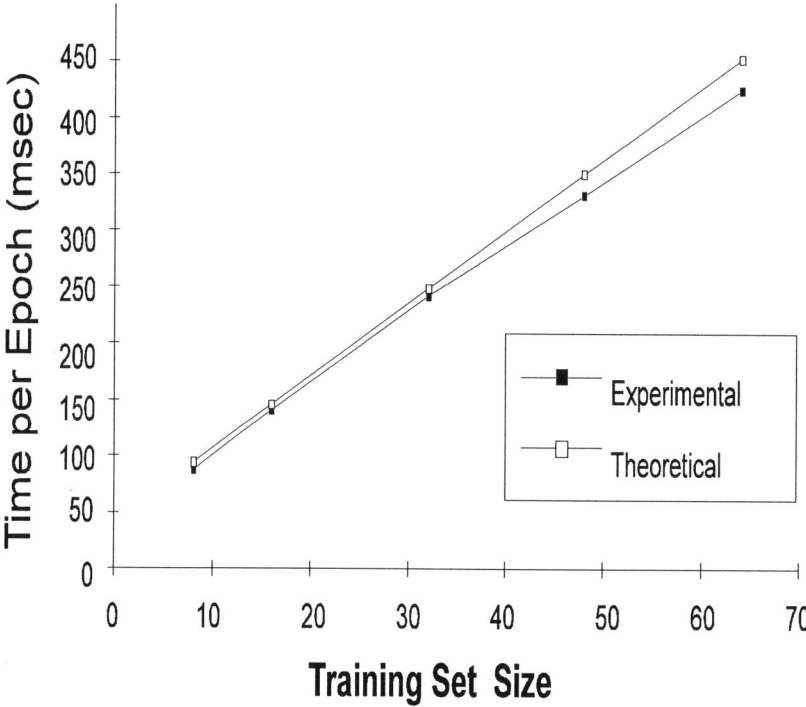

Figure 4.2: Theoretical and experimental time per epoch for the 64-6-64 encoder as a function of training set size.

(within 5%) thus confirming our analysis[1].

---

[1]The slight increase in the theoretically calculated epoch time over the experimental values is mainly due to the use of two dimensional array timings for the machine operations in the theoretical calculations where as in the experimental case the compiler would use both two dimensional and single dimensional operations as appropriate.

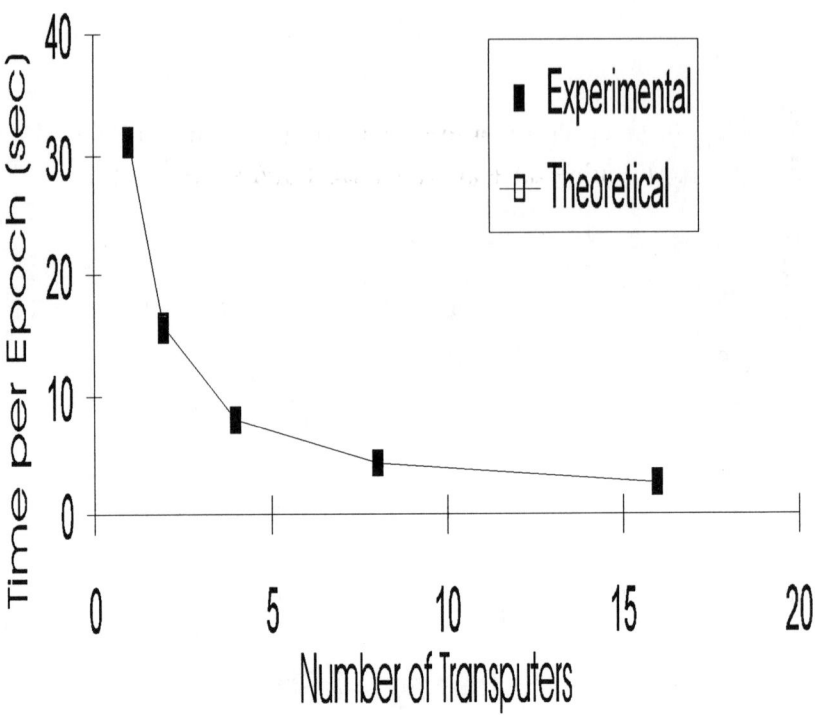

Figure 4.3: Theoretical and experimental time per epoch as a function of the number of transputers for the 256-8-256 encoder.

4.2.1 Speedup

Speedup (S) is the ratio between the epoch time on single processor, and that on a multiprocessor network. The time for an epoch on single processor can be obtained by putting $n = 1$ and ignoring the communication time component in Eq. (4.3).

$$S = \frac{P_{tot}T_c + N_{wb}(t_m + t_a + 2t_{ass} + t_{aa})}{T_{epoch}} \qquad (4.4)$$

The maximum achievable speedup on a fixed size transputer network can be found by letting P_{tot} in Eq. (4.4) become very large. When $P_{tot} \rightarrow \infty$ the speedup becomes,

$$\lim_{P_{tot} \rightarrow \infty} S = n.$$

Thus for neural networks with very large training set the achievable speedup is proportional to the size of the transputer network. Eq. (4.4) can be used calculate the speedup for any neural network as long its size (i.e. number of weights and neurons) and the training set size are given[2]. Fig. (4.4) shows the speedup versus number of transputers for different training set sizes for a 256-8-256 network. It can be seen from the figure that linear speedup occurs only when the training set size is much larger than the square of the number of processors (i.e. $P_{tot} > n^2$).

In the literature on parallel implementation connection updates per second (CUPS) is often used as a performance measure.

The connection updates per second (CUPS) can be calculated from T_{epoch} as[3].

$$CUPS = \frac{N_{wb} P_{tot}}{T_{epoch}}$$

Fig. (4.5) shows the theoretically calculated performance of parallel implementation in terms of CUPS for the 256-8-256 encoder. For the purpose of comparison the performance of a single

---

[2]T_c in Eq. (4.4) also depends on the number of weights and neurons in the network.

[3]The definition for CUPS used here is same as that in [53]

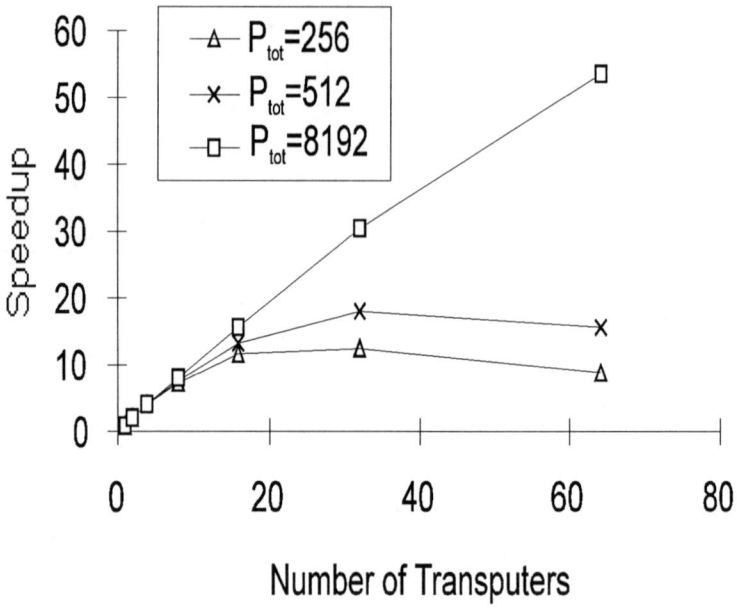

Figure 4.4: Speedup as a function of number of transputers for a 256-8-256 network.

486-33 is also shown. Such comparison are useful in determining the worth of parallel implementations [54]. As can be seen from Fig. (4.5), the benefits of parallelism become apparent only when the number of transputers exceed 7. In many real world situations the benefit has to be weighed against the cost ($) of parallelism[4].

For the specific case of equal distribution some more questions can be answered using the analytical model. For example, we can

---

[4]Currently Transputers cost more than 486 processors.

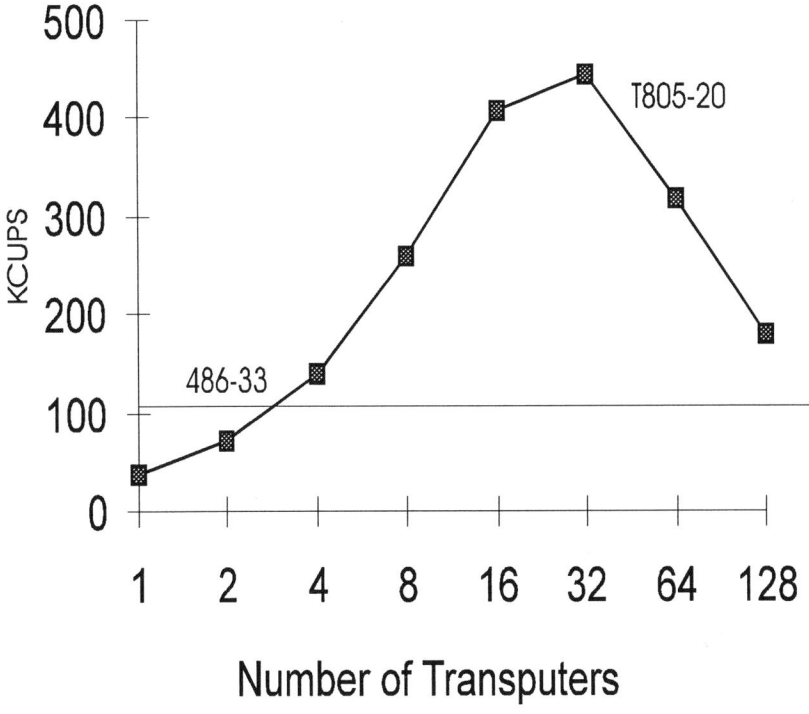

Figure 4.5: Performance (CUPS) as a function of number of transputers for the 256-8-256 encoder.

find the optimal number of processors needed for a given problem. The *optimal* implied here refers to the case when the resulting Time per Epoch is minimum.

4.3 Optimal Number of Transputers Needed for the Case of Equal Distribution

From the mathematical analysis above, it can be seen that the time per epoch Eq. (4.3) for the case when the patterns are equally distributed is given by

$$
\begin{aligned}
T_{epoch} = \ & 2(n-1)[T_{init} + rdN_{wb}] \\
& + \frac{P_{tot}T_c}{n} + (n-1)t_{aas} \\
& + N_{wb}[2t_{ass} + t_m + t_a + t_{aa} + (n-1)t_{aas}] \quad (4.5)
\end{aligned}
$$

Assuming that we are given the size of the neural network and the number of training patterns, based on the above equation, the optimal number of transputers needed for minimum time per epoch can be solved by taking the derivative of the above equation with respect to n and setting it to zero. Noting that T_c is independent of n, the derivative of the above equation is

$$
\frac{\partial T_{epoch}}{\partial n} = 2T_{init} + 2rdN_{wb} - \frac{P_{tot}T_c}{n^2} + N_{wb}t_{aas} + t_{aas} \quad (4.6)
$$

Setting it to zero and solving for n, we have the optimal value for the number of transputers needed for minimum time per epoch is

$$n^* = \sqrt{\frac{P_{tot}T_c}{2T_{init} + (2rd + t_{aas})N_{wb} + t_{aas}}} \qquad (4.7)$$

It can be observed from Eq. (4.7) that the optimal number of transputers needed depends mainly on the number of training patterns present, P_{tot}, the size of the network N_{tot}, and the communication time involved during the transfer of weight changes, $2rdN_{wb}$.

4.4 Cost Benefits Analysis of Adding Additional Processors

From the results in Section 4.3, we note that given a certain number of training patterns and a certain network size, there is always an optimal number of transputers needed for minimizing the time per epoch during training. This result implies that we cannot go on increasing the number of transputers in order to achieve speed up. There will always come a time when the communication overhead will actually exceed the computation time causing it to be inefficient to run on a bigger transputer array. The degradation in performance (D) of not using the optimum number of transputers is given by the following equation (expressed in percentage and normalized to the optimal time per epoch for each encoder configuration):

$$D = \frac{T^n_{epoch} - T^{n^*}_{epoch}}{T^{n^*}_{epoch}}\% \qquad (4.8)$$

where T_{epoch}^{n} is the training time per epoch achieved using n number of transputers, $T_{epoch}^{n^*}$ is the training time per epoch achieved using the optimal number of transputers n^* i.e. when the training time per epoch is minimized

Fig. (4.6) shows the percentage degradation in the time per epoch versus the number of transputers.

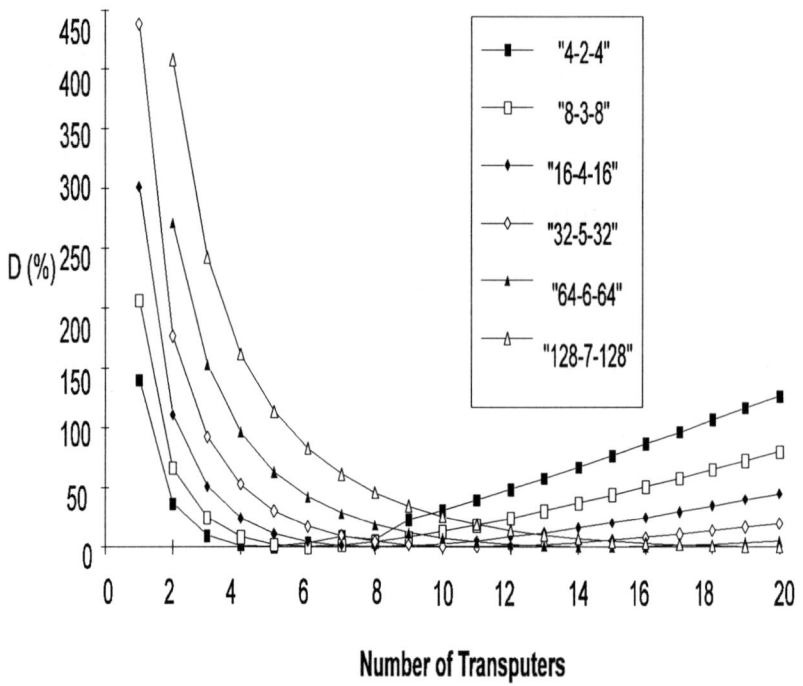

Figure 4.6: Graph showing the percentage decrease in the time per epoch due to a variation to the number of transputers used.

In this figure, encoders of size ranging from 4-2-4 to 128-7-128 are shown. It can be observed from this figure that the slope of the curve is steep before it hits the optimal transputer value and shallow after this optimal value. This observation stresses that the timing advantage gained from adding extra transputers is significant if the current transputer array has a number of transputers smaller than this optimal value. Due to the steep change in gradient before the optimal transputer value, a large improvement in the time per epoch is expected for every additional transputer added to the pipeline-ring system.

The results obtained from Section 4.3 can also be used to find out the percentage advantage of the time per epoch as compared to the 486-33PC. Table (4.1) shows the tabulated comparison for various encoder sizes between the 486-33 PC and the pipeline-ring configuration. In each of the pipeline-ring timing, only the time per epoch with the optimal number of transputers is used.

From the table, it is clearly seen that the computation efficiency of transputers only comes into the picture when the problem size gets bigger.

To summarize, this chapter has provided a mathematical expression for the time per epoch of a set of transputers interconnected together to form a pipeline-ring topology and when the patterns are distributed equally. Based on the analytical model, we can predict accurately the effect of parallelism and the timing advantage obtained from it. The analysis also helps us to give a theoretical optimal number of processors needed for a given neural network problem. This optimal number of processors will enable us to predict the best time per epoch achievable from the pipeline-ring topology. If the user finds that this optimal time is not satisfactory or if the optimal number of transputers needed is too large,

Table 4.1: Table showing the time per epoch comparison between the 486-33PC and the pipeline-ring topology configuration of transputers.

Encoder Size	486-33PC Time per Epoch (msec)	Pipeline-ring Time per Epoch (msec)	Percentage Improvement over 486-33PC(%)	Optimal Number of T805s
4-2-4	1.981	2.337	-17.97	4.17
8-3-8	7.908	7.699	2.64	5.50
16-4-16	32.355	24.786	23.39	7.32
32-5-32	134.322	79.014	41.18	9.88
64-6-64	571.171	250.032	56.22	13.52

the user can resort to other means of parallelism. All these can be done without the need to conduct the actual experiments in a transputer array.

Chapter 5

Optimization Model for Unequal Distribution of Patterns in a Homogeneous Array of Transputers

In the previous chapter, we considered the case where the training set is assumed to be divisible by the number of processors present and each processor is allocated patterns equally. However, in many cases, the number of training patterns present in the network may not be fully divisible by the number of processors present. For such cases, unequal distribution is necessary. In this chapter, we analyze how this distribution can be done in order to minimize the training time per epoch. In the previous sections, we have concentrated in getting a model for T_{epoch} only without including the constraints on the memory. In a realistic problem, especially when one is dealing with unequal distribution of patterns, memory constraints of the processors also have to be taken into account. i.e. the available memory in each processor is limited and hence one can not allocate any number of patterns to that processor. When this constraint is imposed on the problem of finding optimal pattern allocation schemes (optimum in the sense of minimum time per epoch) for unequal distribution, it is found that such an analysis results in solving a mixed integer programming problem. Using this approach, it is found that even when the training set size is an integer multiple of the number of processors, allocating equal number of patterns to each processor is not necessarily the optimal way to distribute the patterns amongst the processors especially when the size of the transputer network is large.

The model for T_{epoch} developed in Chapter 3 covers the general case for unequal pattern distribution. It is worth while to recollect Eq. (3.7) which is given below.

$$T_{epoch} = T_{con}^{\Delta wo} + P_{(n)}T_c + w_n^r + T_{con}^{\Delta w} + T_{con}^w + n[T_{init} + T_{trf}] \quad (5.1)$$

It can be seen from the above equation that the time per epoch depends on $P_{(n)}$ and w_n^r. The other terms in Eq. (3.7) are constants for a given processor network and neural network. But the waiting time w_n^r depends on the time spent by the $(n-1)^{th}$ processor on computations and waiting i.e. P_{n-1}, w_{n-1}^r and w_{n-1}^s, which in turn depends on P_{n-2}, w_{n-2}^r and w_{n-2}^s and so on. Thus minimizing T_{epoch} requires choosing $P_1, P_2, .., P_n$ in such a way that $P_nT_{cn} + w_n^r$ is made as low as possible.

This minimization problem can be posed in a compact matrix form as

$$\min_{P_i} \ \{ \ T_{epoch} = CP + w_n^r \ \} \quad (5.2)$$

where C and P are defined as

$$C = \begin{bmatrix} 0 & .. & . & T_{cn} \end{bmatrix}_{1 \times n} \quad P = \begin{bmatrix} P_1 \\ . \\ . \\ . \\ P_n \end{bmatrix}_{n \times 1}$$

How to optimally choose $P_1, P_2, .., P_n$ so as to minimize T_{epoch} and the constraints involved in the optimization are analyzed in the next section.

5.1 Constraints for Optimization

The algorithm to optimally distribute the training set must minimize T_{epoch} without violating any constraints imposed by the processor network. The various constraints that must be satisfied are described below.

5.1.1 Synchronized inter process communication

Because interprocess communications are always synchronized, a processor cannot send data to another unless the other processor is ready to receive it. In Fig. (3.1) which is reproduced here as Fig. (5.1), the horizontal arrow ($\rightarrow$) between any two processors indicates the time at which data transfer between the two processors is complete. Between processors 1 and 2 it can be seen from the timing diagram that the time at which processor 1 completes sending data to processor 2 is given by

$$T_1 = T_{con}^{\Delta w_0} + P_1 T_c + w_1^s + T_{init} + T_{trf}^{\Delta w} \tag{5.3}$$

and the time at which processor 2 completes receiving data from processor 1 is

$$T_2 = T_{con}^{\Delta w_0} + P_2 T_c + w_2^r + T_{init} + T_{trf}^{\Delta w} \tag{5.4}$$

These two times must be equal as interprocess communications are always synchronized. Hence between processors 1 and 2 the following equation must hold.

$$T_{con}^{\Delta w_0} + P_1 T_c + w_1^s = T_{con}^{\Delta w_0} + P_2 T_c + w_2^r \tag{5.5}$$

and between any two processors, $j - 1$ and j, where $3 \le j \le n$ we have

$$T_{con}^{\Delta w_0} + P_{j-1} T_c + w_{j-1}^r + T_{init} + T_{trf}^{\Delta w} + T_{con}^{\Delta w} + w_{j-1}^s = T_{con}^{\Delta w_0} + P_{(j)} T_c + w_j^r \tag{5.6}$$

The number of patterns assigned to the various processors must be non-negative and their sum must equal the training set size P_{tot}; i.e.

$$\sum_{i=1}^{n} P_i = P_{tot} \tag{5.7}$$

and P_i must be non-negative integers.

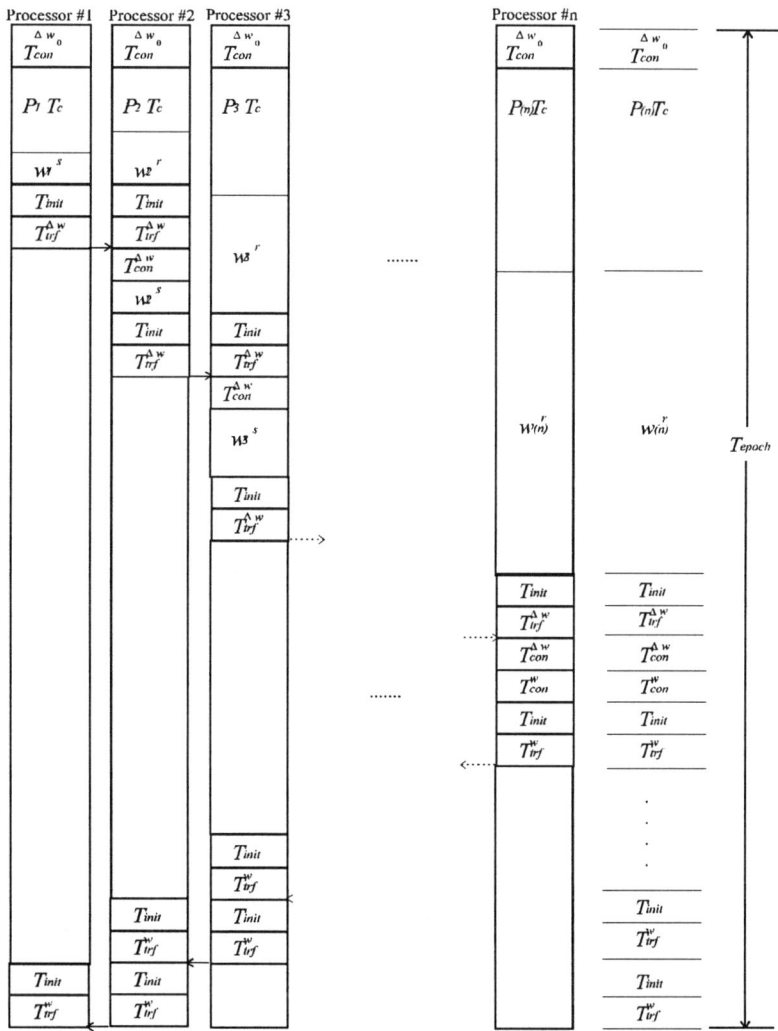

Figure 5.1: Timing Diagram for an epoch for a homogeneous array of transputers

This constraint can be put in a compact matrix form as:

$$AP + W = B$$

where

$$A = \begin{bmatrix} T_c & -T_c & 0 & .. & 0 \\ 0 & T_c & -T_c & .. & 0 \\ . & . & . & .. & . \\ . & . & . & .. & . \\ 0 & . & . & T_c & -T_c \\ 1 & 1 & 1 & 1 & 1 \end{bmatrix}_{n \times n}$$

$$B = \begin{bmatrix} 0 \\ -(T_{init} + T_{trf}^{\Delta w} + T_{con}^{\Delta w}) \\ . \\ . \\ -(T_{init} + T_{trf}^{\Delta w} + T_{con}^{\Delta w}) \\ P_{tot} \end{bmatrix}_{n \times 1}$$

$$W = \begin{bmatrix} (w_1^s - w_2^r) \\ (w_2^s + w_2^r - w_3^r) \\ (w_3^s + w_3^r - w_4^r) \\ . \\ . \\ (w_{n-1}^s + w_{n-1}^r - w_n^r) \\ 0 \end{bmatrix}_{n \times 1}$$

$$P = \begin{bmatrix} P_1 \\ \cdot \\ \cdot \\ P_n \end{bmatrix}_{n \times 1} \qquad P \geq 0 \ \ and \ \ P_i \ \ are \ \ integers.$$

5.1.2 Processor waiting times must be non-negative

The time spent by the processors on waiting to receive and send data must be non-negative.

$$
\begin{aligned}
w_i^r &\geq 0, \ and \\
w_i^s &\geq 0, \ \ 1 \leq i \leq n.
\end{aligned}
\tag{5.8}
$$

In matrix form, the above constraint becomes

$$W^s = \begin{bmatrix} w_1^s \\ \cdot \\ w_{(n-1)}^s \\ 0 \end{bmatrix}_{n \times 1} \qquad W^r = \begin{bmatrix} 0 \\ w_2^r \\ \cdot \\ w_{(n)}^r \end{bmatrix}_{n \times 1} \qquad and \ \ W^s \geq 0 \ \ W^r \geq 0$$

5.1.3 Memory limits in the processors

In training-set parallelism each processor has to keep a complete copy of the neural network as well as the patterns that are allocated to it. Since each transputer has a fixed external memory, (Eg. 1

Mbytes, 2 Mbytes, 4 Mbytes, etc) there is a limit to the number of network parameters (weights, activations, etc) and training patterns that a processor can hold. In this section we examine the memory requirement for the processors in the pipelined ring network for implementing the parallelized backpropagation algorithm.

Storing the network parameters

If r is the number of bytes required to store a floating point number, then the amount of memory required to store all the weights and biases in a neural network is

$$r(N_w + N_{tot}) \qquad (5.9)$$

where N_w is the number of weights in the network and N_{tot} is the number of neurons (excluding neurons in the input layer) in the network. N_{tot} also corresponds to the number of biases in the neural network.

Storing the training set

Memory is needed to store the training patterns allocated to a processor. Corresponding to each input (training) pattern there is a desired output pattern. If N_1 and N_l are the number of neurons in the input and output layer respectively, to store each training pattern $r(N_1 + N_l)$ amount of memory is required. The total memory required to store all the P_i patterns allocated to a processor i will be

$$P_i \, r(N_1 + N_l) \qquad (5.10)$$

Forward pass of the backpropagation

The forward pass of the backpropagation requires the output value at every neuron and this in turn requires the input value at every neuron. Since there are N_{tot} neurons in the network the amount of memory required to store these will be

$$r(N_{tot} + N_{tot}). \qquad (5.11)$$

Backward pass of the backpropagation

During the backward pass of the backpropagation each processor has to store the weight changes, bias changes and at each neuron the errors ('deltas'). The memory required for all these will be

$$r(N_w + N_{tot} + N_{tot}). \qquad (5.12)$$

If momentum term is used in the backpropagation then the weights and biases of the previous epoch need to be saved. This requires a memory of

$$r(N_w + N_{tot}) \qquad (5.13)$$

but only for the last processor as the weights are updated only in the last processor. In our implementation we used the momentum term when updating the weights in the last transputer.

Overall memory requirement

The overall memory requirement for all the transputers except the last one is obtained by adding the expressions (5.9) to (5.12) which results in

$$r[2N_w + 5N_{tot} + P_i(N_1 + N_l)]. \tag{5.14}$$

For the last transputer in the pipelined ring, expression (5.13) is added to expression (5.14) to give the memory requirement as

$$r[3N_w + 6N_{tot} + P_n(N_1 + N_l)]. \tag{5.15}$$

The external memory M_i in the transputers should be greater than the above memory requirements i.e.

$$M_i > r(2N_w + 5N_{tot} + P_i(N_1 + N_l)), \quad 1 \le i \le n - 1 \tag{5.16}$$

$$M_i > r(3N_w + 6N_{tot} + P_i(N_1 + N_l)) \quad i = n. \tag{5.17}$$

This constraint can be compactly put in matrix form as:

$$E < F \tag{5.18}$$

where

$$E = \begin{bmatrix} r(2N_w + 5N_{tot} + P_1(N_1 + N_l)) \\ r(2N_w + 5N_{tot} + P_2(N_1 + N_l)) \\ . \\ . \\ r(3N_w + 6N_{tot} + P_n(N_1 + N_l)) \end{bmatrix}_{n \times n}$$

and

$$F = \begin{bmatrix} M_i \\ . \\ . \\ M_i \end{bmatrix}_{n \times 1}$$

5.2 Optimal Pattern Distribution

Eqs. (5.5) - (5.17) are constraints that must be satisfied by the optimal distribution algorithm. Finding the optimal distribution of P_i to minimize the epoch time T_{epoch} subject to the constraints given by Eqs. (5.5) - (5.17) is a mixed integer programming problem (mixed because P_i's must be integers but the w_i^T's and w_i^s's can be fractions) as described below.

$$\min_{P_i} \ \{ \ T_{epoch} = CP \quad + \quad w_n^r \ \} \qquad (5.19)$$

$$Subject\ to \quad AP \quad + \quad W = B \qquad (5.20)$$

$$E \quad < \quad F \qquad (5.21)$$

$$\sum_{i=1}^{n} P_i \quad = \quad P_{tot} \qquad (5.22)$$

$$P \geq 0, \quad W^s \geq 0, \quad W^r \geq 0 \qquad (5.23)$$

$$and\ the\ elements\ of\ P\ are\ integers. \qquad (5.24)$$

where $A, B, C, E, F, P, W, W^s, W^r$ have been defined earlier.

Minimization of Eq. (5.19) can be done using such method as the Branch and Bound algorithm [55]. The general idea of branch and bound is to first solve the problem without the integer restrictions. This will reduce the problem into a linear programming problem (first node). If in the optimal solution an integer variable x_k takes a non integer value x_k^* two linear programming subproblems are created by branching the first sub-problem imposing an extra constraint $x_k \leq [x_k^*]$, while the second sub-problem imposes an extra constraint $x_k \geq [x_k^*]$. These two linear programming problems will now be solved. The method of branching continues until an integer solution (bound) is found. This solution is then compared with other similarly found integer solutions. The method of branching and bounding continues until the optimal solution is found. Appendix C shows a detailed explanation of the branch and bound method of solving the mixed integer programming problem. A worked example illustrating the method is also included.

5.3   Validation of the Pattern Optimization Model

The above optimization model has been derived based on theoretical models. In this section, detailed sets of experimental results are presented so that they can be compared with the theoretical results. For our comparison, we measure the time taken per epoch for the 256-8-256 encoder described earlier. A set of T805-20 is used for all the experiments that are conducted in this section. The time taken for equal distribution with a varying number of transputers is tabulated. The number of transputers used are 1, 2, 4, 8, 16. The experimental along with theoretical results are tabulated in Table (5.1). Fig. (5.2) shows the graphical representation of the Table (5.1). It can be observed that in all cases, the experimental time taken is always lower than that of the theoretical time taken. This is generally true as two dimension arrays are used in the measurements of the basic arithmetic operations. More details can be found in Section 3.4 of Chapter 3. Other than this, the difference between the two sets of values are relatively small which verifies the theoretical results established earlier.

Table 5.1: Table showing the comparison between the experimental and theoretical time per epoch for the 256-8-256 encoder (equal distribution)

Number of Transputers	1	2	4	8	16
Experimental Time per Epoch(sec)	30.892	15.510	7.756	4.108	2.506
Theoretical Time per Epoch(sec)	31.289	15.735	8.027	4.313	2.735

Another set of experiments is conducted and this time the comparison is conducted for optimal pattern distribution. The optimal pattern distribution for the 256-8-256 encoder problem is used for processor network sizes of 1, 2, 4, 8, 10, 12, 14 and 16. Table (5.2) shows the tabulated theoretical and experimental results. Fig. (5.3) shows the graphical representation of Table (5.2). The small difference in the two sets of results indicates that the theoretical results gives a good prediction of the experimental results.

We can make use of the results in Table (5.2) to obtain the speedup, S as defined in Section 4.2.1 of Chapter 4. The higher the speedup value for a given number of processors implies that the processors are better utilized. It is noted that the single transputer implementation of the backpropagation algorithm is different from the parallel version because there is no message passing involved. Fig. (5.4) shows the speedup calculated from Table (5.2). The

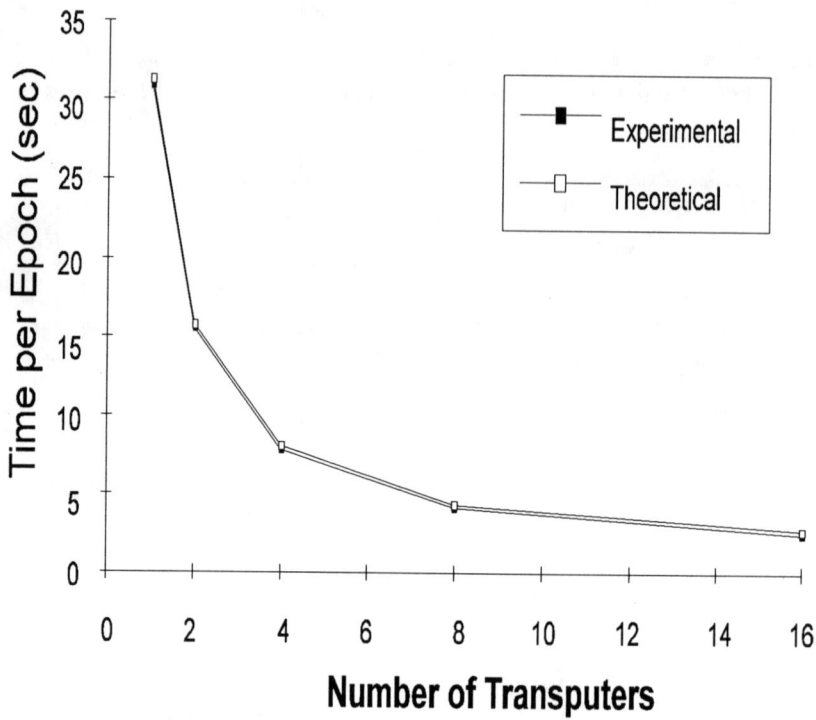

Figure 5.2: Graph showing the comparison between experimental and theoretical time per epoch for the 256-8-256 encoder (equal distribution)

tabulated results of the speedups are found in Table (5.3).

Table 5.2: Table showing the comparison between experimental and theoretical time per epoch for the 256-8-256 encoder (optimal distribution)

Number of Transput-ers	1	2	4	8	10	12	14	16
Experimental Time per Epoch (sec)	30.89	15.51	7.756	4.059	3.284	2.876	2.542	2.312
Theoretical Time per Epoch (sec)	31.28	15.73	8.027	4.276	3.414	3.085	2.765	2.551

Table 5.3: Table showing the comparison between experimental and theoretical speedups for the 256-8-256 encoder

Number of Transput-ers	1	2	4	8	10	12	14	16
Experimental Speedup	1.000	1.991	3.983	7.611	9.407	10.74	12.15	13.36
Theoretical Speedup	1.000	1.988	3.898	7.317	9.165	10.14	11.31	12.26

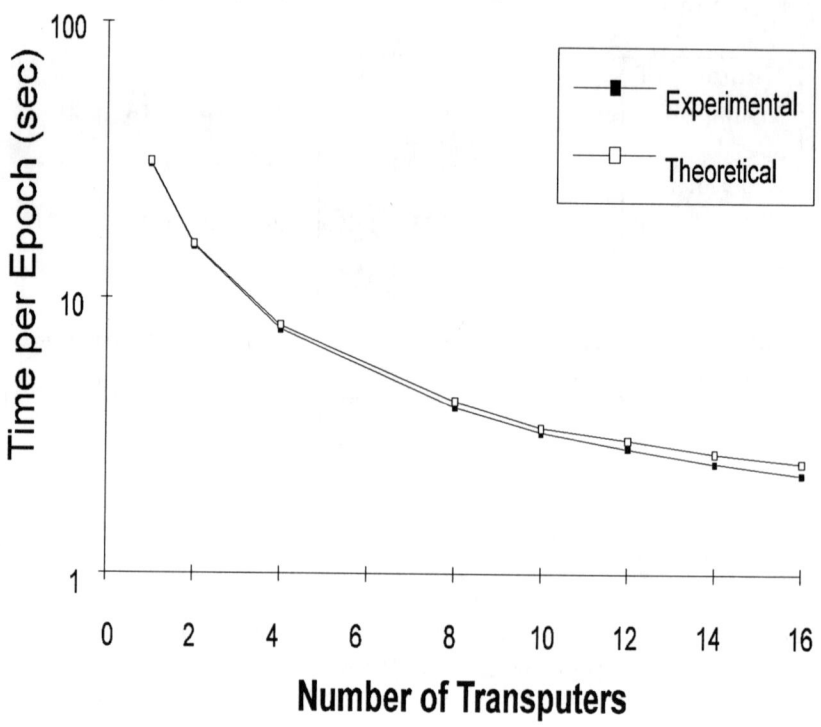

Figure 5.3: Graph showing the comparison between experimental and theoretical time per epoch for the 256-8-256 encoder problem (optimal distributed)

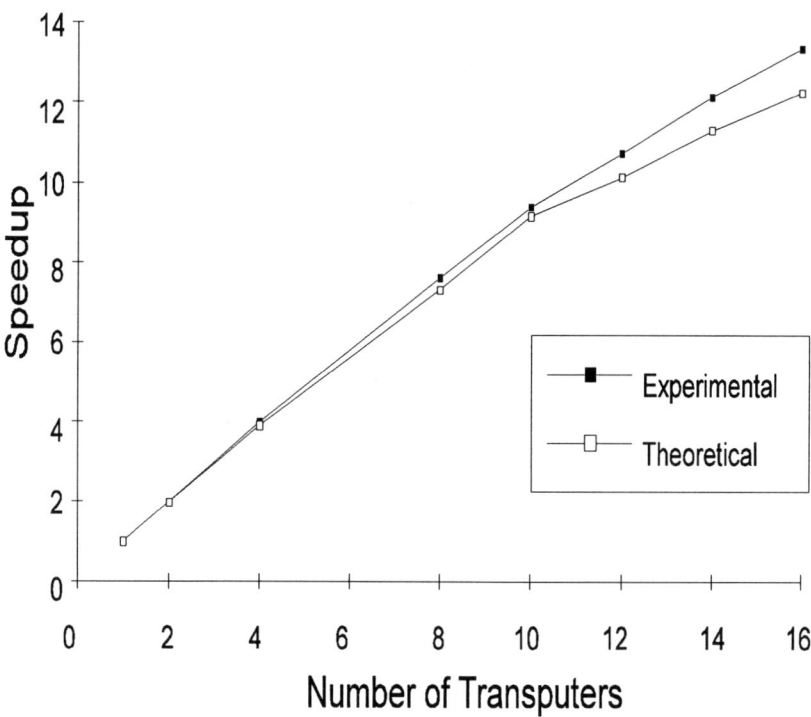

Figure 5.4: Graph showing the comparison between experimental and theoretical speedups for the 256-8-256 encoder

5.4   Experimental Results for Benchmark problems

5.4.1   An overview of the experiments conducted

This section gives examples on how the analysis in the preceding sections can be applied for a homogeneous processor array. That is to say, each processor in the pipelined ring runs at the same processing speed and have the same memory size. Following experiments are conducted and the results are compared with the theoretical results. The objectives of these experiments can be summarized as:

- To investigate whether equal distribution of patterns is always the best way to distribute a set of patterns when the total number of patterns is an integer multiple of the number of processors present

- To show how we can distribute patterns optimally when the total number of training patterns present is not an integer multiple of the number of processors present

- To make use of the results in previous sections to locate processors which are not required in the processor network (surplus processors)

- To find out the optimal number of processors needed to obtain minimum time per epoch

For all the experiments conducted here, transputer model T805-20 is used. The mixed integer programming is performed using the NAG FORTRAN Library Routine H02BBF [56]. This routine uses the Branch and Bound method to solve a mixed integer programming problem. The simulations are performed on the three benchmark problems described in Chapter 2: The N-X-N encoder problem [48], the sonar problem [50] and the NETTALK problem [49]. Experimental results as well as theoretical results are both presented to verify the formulations developed in the earlier sections.

5.4.2   Is equal distribution optimal ?

When the number of training patterns used are wholly divisible by the number of processors present, the usual approach is to divide the patterns equally among the processors. It can be seen from Fig. (3.1) that doing so will result in time wastage. This is due to the presence of the waiting time involved. In order to make use of these waiting times to lower down the training time per epoch, the analysis in the previous sections can be used to fill in as many patterns as possible.

For example, consider the 256-8-256 encoder on a processor network with 16 transputers. Using the mixed integer programming technique developed in earlier in this chapter, the optimal distribution of patterns is found to be 14, 14, 14, 15, 15, 15, 16, 16, 16, 16, 17, 17, 17, 18, 18, 18 and the corresponding theoretical epoch time is 2551 msecs. This is equivalent to 437.54 KCUPS. Using equal distribution (16 patterns per processor) the theoretical epoch time

becomes 2735 msecs (408.10 KCUPS). This means equal allocation is not the best way to distribute the patterns amongst the
processors. To show the difference in time for equal and optimal
distributions, a series of experiments are conducted. For each of
the experiments described, the same 256-8-256 encoder problem is
used, the only difference in the experiment is the number of transputers used. The number of transputers used are 1, 2, 4, 8 and 16
respectively. It is noted that in all cases, the number of transputers
chosen is divisible by the number of training patterns. Fig. (5.5)
shows the theoretical results for the time per epoch for both equal
distributed and optimal distributed curves. Table (5.4) shows the
tabulated results of Fig. (5.5). It can be seen from the timing of
this table that when the number of transputers used is greater or
equal to eight, there is a difference in the pattern distribution between equal and the optimal case. The difference in the time tends
to increase when the number of processors used increases.

Table 5.4: Table showing the difference in time per epoch between
equal and optimal distribution for 256-8-256 encoder

Number of Transputers	1	2	4	8	16
Equal Distri- bution Time per Epoch(sec)	31.289	15.735	8.027	4.313	2.735
Optimal Distribu- tion Time per Epoch(sec)	31.289	15.735	8.027	4.276	2.551

It is noted that for the sixteen transputers case, there is only a

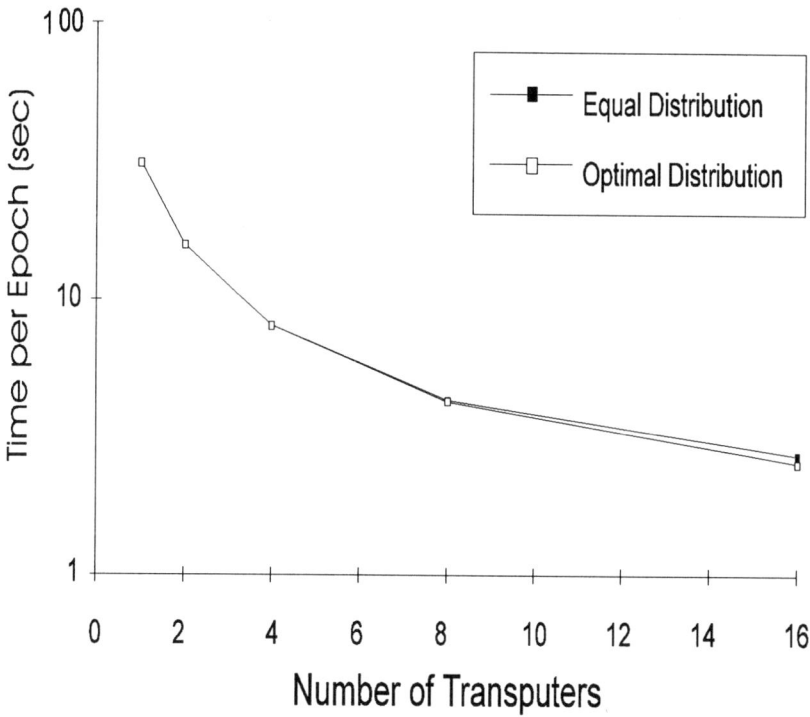

Figure 5.5: Comparison between optimal and equal distribution for 256-8-256 encoder

small difference in the time per epoch between the optimal distribution and equal distribution for this 256-8-256 encoder problem. The difference being 7.2%. This slight difference is because the number of transputers used in this processor network is small compared with the optimal number of transputers needed for this network. It is worked out theoretically that the optimal number of transputers needed for this 256-8-256 encoder is 34. To have a meaningful value for comparison, 32 transputers are used in the following discussion. 32 transputers are chosen so that the training patterns

can be divided equally among the transputers. Since the number
of transputers currently available is less than 32, only simulation
results can be provided here. For equal distribution, the theoretical time per epoch taken is 2504.389 msec (445.68 KCUPS). The
theoretical time per epoch for optimal distribution obtained from
the NAG routine is 2110.403 msec (528.88 KCUPS). A difference of
almost 19% is obtained from these timings. This shows clearly that
it is much beneficial to obtain the optimal pattern distribution for
bigger processor networks as more time can be saved per training
epoch.

Another example deals with the sonar problem as described
in [50]. This sonar target classification problem uses 192 training
patterns. Each neural network has 60 input nodes, 24 hidden nodes
and 2 output nodes making a total connection of 1514 weights. A
total of twelve T805-20 transputers are chosen because this number
is divisible by the number of training patterns present. This aids
in the comparison between equal distribution and optimal distribution.

For equal pattern distribution, the theoretical time per epoch
obtained is 794.79 msec and the actual time taken is 620 msec
(468.85 KCUPS). The optimal distribution for this twelve transputer configuration is 14, 14, 15, 15, 15, 16, 16, 17, 17, 17, 18, 18.
The theoretical time per epoch is found to be 751.50 msec and the
experimental result is 581 msec (500.32 KCUPS). Fig. (5.6) shows
the theoretical time per epoch between equal and optimal distributions for various processors sizes. Table (5.5) shows the tabulated
results of Fig. (5.6).

For this sonar problem, the optimal pattern distribution starts

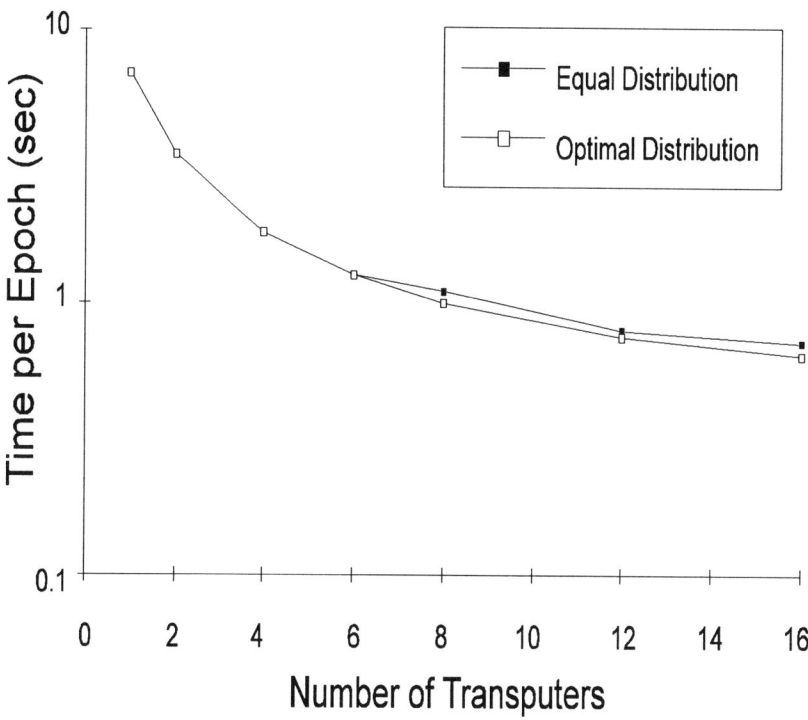

Figure 5.6: Graphical comparison between equal and optimal distribution for the sonar problem

to deviate from equal distribution when the number of processors exceeds 4. This number is expected to be smaller from the 256-8-256 encoder case as this is a smaller benchmark problem. Generally, the bigger the network size, the larger is the number of processors needed to result in any difference in the pattern distribution between optimal distribution case and equal distribution case.

Table 5.5: Table showing the comparison between equal and optimal distribution for the sonar problem

Number of Transputers	1	2	4	6	8	12	16
Equal Distribution Time per Epoch(sec)	6.978	3.521	1.818	1.273	1.107	0.795	0.717
Optimal Distribution Time per Epoch(sec)	6.978	3.521	1.818	1.269	1.000	0.752	0.645

5.4.3   Examples when the total number of training patterns is not an integer multiple of the number of processors present

When the total number of training patterns is not divisible by the number of transputers present, a heuristic approach [15] can be as follows. Starting from the last transputer to the first, assign the number of training patterns in each transputer as $(P_{tot}/n)+1$ for the first $(P_{tot} \bmod n)$ transputers and P_{tot}/n patterns for the remaining transputers. P_{tot}/n refers to the integer division operator between P_{tot} and n. However due to the waiting time involved because of the need for synchronization, better pattern distributions could be achieved by using the optimization method discussed earlier. This section shows how we can optimally distribute patterns that are not integer multiples of the number of transputers.

Consider the 256-8-256 encoder with 256 Input/Output training patterns and 4096 connections including biases. In a processor network with ten T805-20 processors, equal distribution of patterns would result in 25.6 patterns per processor. This calls for some form of unequal distribution. Using the mixed integer optimization method we found that the optimal distribution of patterns is 24, 25, 25, 25, 25, 26, 26, 26, 27, 27. The resulting epoch time from theoretical calculation is 3552.40 msecs. The experimental time found is 3339 msecs per epoch (314.04 KCUPS). Any other way of allocation of input patterns would produce higher epoch time. As an example, consider the following two straight forward allocation strategies.

i) Allocate the first 4 processors with 25 patterns and the last 6 processors with 26 patterns. In this case the theoretical time per epoch becomes 3564.10 msecs. The corresponding experimental result is 3354.7 msecs.

ii) Allocate the first 9 processors with 26 patterns and the last with 22 patterns. The theoretical time per epoch then becomes 3638.43 msecs and the experimental time per epoch is 3473 msecs.

The experiments above clearly show that Eq. (5.19) can be used to find better time per epoch as compared to the intuitive way of allocating patterns to processors having unequal pattern distribution. Although the difference in timing between the intuitive way of distributing and the optimal way of distributing is not large, it can be easily predicted that for larger networks, more time savings can be expected. Assuming that 34 transputers are used instead of 20, the theoretical optimal time obtained is 1702.7 msec and the intuitive distributions gives a time per epoch of 2073.3 msec. This results in a percentage change of about 18% over the intuitive distribution case.

Next, consider the NETTALK text to phoneme problem on a neural network with 13,826 connections (203 by 60 by 26) and with 12022 input patterns. In a processor network with twenty T805-25 transputers equal division of patterns would result in 601.1 patterns/processor. Using the mixed integer optimization method we calculated the optimal distribution of patterns to be 597, 597, 598, 598, 599, 599, 600, 600, 600, 601, 601, 602, 602, 603, 603, 604, 604, 604, 605, 605. The resulting time for an epoch is 136486.49 msecs which is 1217.82 KCUPS. This will be the minimum time for an epoch on this processor network for this problem. Any other allocation of patterns should result in a higher epoch time. Again as an example consider the following two straight forward allocation strategies.

i) Allocate the first 18 processors with 601 patterns and the last 2 processors with 602 patterns. In this case the time per epoch becomes 137214.35 msecs (1211.36 KCUPS).

ii) Allocate the first 9 processors with 600 patterns and last 11 processors with 602 patterns. The time per epoch then becomes 136990.59 msecs (1213.34 KCUPS).

All these examples indicate that for bigger processor networks, the intuitive way of allocating patterns among the processors is not the best way to give a low time per epoch. And since the number of training epochs normally increases when the complexity of the problem increases, reducing the training time taken for every epoch will reduce the overall training time.

5.5 Locating Surplus Processors and to Find Out the Optimal Number of Processors Needed to Obtain Minimum Time Per Epoch

Another issue that can be investigated is to make use of the analysis to locate surplus processors in the pipelined ring. A processor is considered a surplus if the mixed integer programming solution allocates zero patterns to it. Such a condition indicates that that processor is not needed in the processor network. Instead, its presence will actually cause the time per epoch to increase. Such a situation occurs when there are too many processors present in the processor network. With the help of this information, it serves as a guide to find out the optimal number of processors needed to minimize the time per epoch. This optimal number of processors will give the least time per epoch for a given neural network problem. For example, given a neural network with 50 processors, the number of surplus processors found could be 10. This indicates that 40 (50 - 10) processors are needed. The optimal number of processors needed is usually close to 40.

As an example, we consider the 16-4-16 encoder problem. In this encoder problem, there are 16 training patterns. This encoder has a total of 148 connection weights. Table (5.7) shows the optimal time per epoch taken for this encoder problem with varying number of transputers used. The transputer model used in this experiment is T805-20. Fig. (5.7) shows the graphical representation of Table (5.7). From this figure, it can be observed that the minimum time per epoch occurs when 10 processors are used. When a processor network has more than this number of processors, the time per epoch will increase. Table (5.6) shows the optimal pattern distributions for each of the transputer configuration used.

Table 5.6: Table showing the optimal pattern distribution for the 16-4-16 encoder

Number of T805s	Pattern Distribution in Each Transputer															
	1	2	3	4	5	6	7	8	9	10	11	12	13	14	15	16
1	16															
2	8	8														
4	4	4	4	4												
6	2	2	3	3	3	3										
8	1	1	2	2	2	2	3	3								
9	1	1	1	2	2	2	2	2	3							
10	1	1	1	1	1	2	2	2	2	3						
11	1	1	1	1	1	1	1	2	2	2	3					
12	0	0	1	1	1	1	1	2	2	2	2	3				
14	0	0	0	0	1	1	1	1	1	2	2	2	2	3		
16	0	0	0	0	0	0	1	1	1	1	1	1	2	2	3	3

From this table, it can be seen that at the optimal number of processors, each processor holds a certain number of patterns. However, when the number of processors used increases to 12 and beyond, the first few processors start to be allocated zero patterns. These processors are known as surplus processors which can be removed from the processor network. Another interesting observation from Table (5.6) is that surplus transputers tend to appear slightly after the optimal number of transputers needed for that encoder problem. This shows that the optimal number of transputers needed for a neural network can be predicted by locating the surplus transputers.

Table 5.7: Table showing the theoretical optimal time per epoch for the 16-4-16 encoder

Number of Transputers	1	2	4	6	8	9	10	11	12	14	16
Theoretical Optimal Time per Epoch (msec)	88.8	47.4	29.2	25.8	23.9	23.9	23.4	24.3	24.7	25.3	27.0

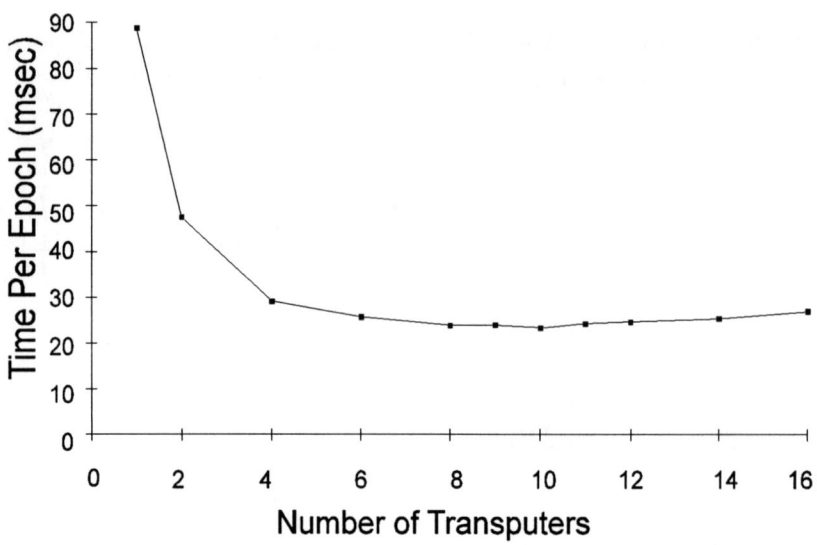

Figure 5.7: Figure showing the theoretical optimal time per epoch for the 16-4-16 encoder

Chapter 6

Optimization Model for Unequal Distribution of Patterns in a Heterogeneous Array of Transputers

The previous chapter discussed the issue of optimally distributing a set of training patterns in a homogeneous array of transputers. In this chapter, we shall extend our study to the case when the transputers used are of different speeds and memory sizes (i.e. heterogeneous processor network) [57]. Although it may seem strange to mix processors of different speeds, it becomes a practical necessity when there are not enough transputers of a single model in the laboratory. We feel that any laboratory inventory on transputers will most likely have some new and some old models of the processor. Using all the available transputers, a processor network may be able to produce shorter epoch times than using only a few transputers of a single model. Analysis shows that in such cases the optimization will tend to find a distribution that will place more training patterns among the faster processors.

In the mathematical formulations of the optimal distribution of a set of patterns among a heterogeneous array of transputers, the approach used is similar to that in Chapter 5. We will first obtain a time per epoch expression from the timing diagram. From the time per epoch expression, we will then proceed to solve for the optimal distribution of training patterns using mixed integer programming.

In a heterogeneous array of transputers connected in a pipeline-ring architecture, the timing diagram for one epoch is shown in Fig. (6.1). In this figure, $P_1 \ldots, P_{(n)}$ are the number of patterns allocated to the processors $1, \ldots n$. $w_1^s, w_2^s, w_2^r, \ldots, w_{n-1}^s, w_{n-1}^r, w_n^r$ are the times spent on waiting to send and receive weight updates and error at each processor.

The various notations used are explained below:

- $T_{con(i)}^{\Delta w_0}$: Time taken by the ith processor for initializing the weight changes.

- $T_{init(i)}$ Time taken for initializing a communication by the ith processor.

- $T_{trf(i)}^{\Delta w}$: Time taken for sending the weight changes by the ith processor.

- $T_{con(i)}^{\Delta w}$: Time for accumulating the weight changes and error by the ith processor.

- $T_{con(n)}^{w}$: Time for updating the weight values in the last pipe processor.

- $T_{trf(i)}^{w}$: Time taken for sending the updated weights by the ith processor.

- T_{ci}: Time taken to perform one forward and backward pass of the BP algorithm by the ith processor.

- P_i: Number of training patterns allocated the ith processor.

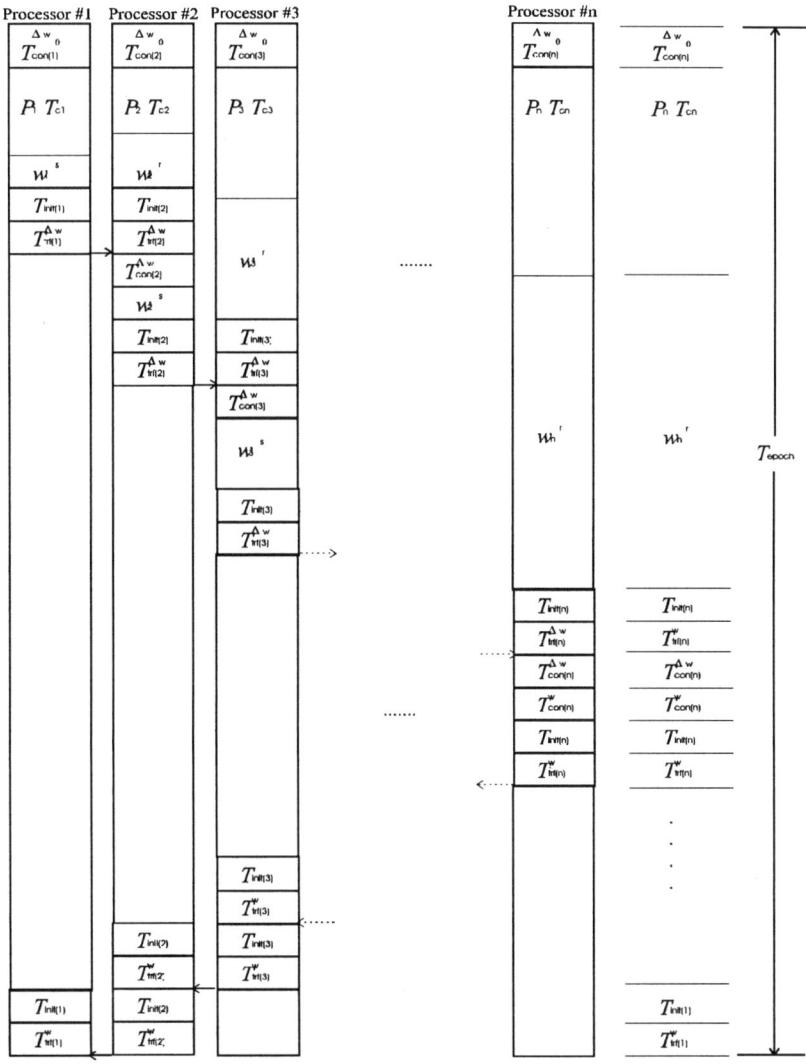

Figure 6.1: Timing diagram for unequal distribution of patterns amongst the processors

- w_i^s: Time spent on waiting by processor i to send weight updates and error to the next pipe processor. The waiting occurs when processor i is ready to send but the next processor is still performing computations and consequently not ready to receive the data.

- w_i^r: Time spent on waiting by processor i to receive weight updates and error from the previous pipe processor. The waiting occurs when processor i is ready to receive but the previous processor is still performing computations and consequently not ready to send the data.

From the timing diagram of Fig. (6.1) the time for an epoch (T_{epoch}) can be written as,

$$T_{epoch} = P_n T_{cn} + w_n^r + T_{con(n)}^{\Delta w_0} + T_{init(n)} + T_{trf(n)}^{\Delta w}$$
$$+ T_{con(n)}^{\Delta w} + T_{con(n)}^w + \sum_{i=2}^{n} [T_{init(i)} + T_{trf(i)}^w] \qquad (6.1)$$

Following similar analysis found in Chapter 5, the above equations can be expressed in a matrix format. By treating the set of constraints as a mixed integer programming problem, the optimal distribution of patterns across this heterogeneous array of processors can be found by solving this set of equations using method like branch and bound [55]. The modified matrix equations are described below.

$$\min_{P_i} \{ CP + w_n^r \} \qquad (6.2)$$

$$Subject\ to\ AP\ +\ W = B \tag{6.3}$$

$$E \leq F \tag{6.4}$$

$$\sum_{i=1}^{n} P_i = P_{tot} \tag{6.5}$$

$$P \geq 0,\ W^s \geq 0,\ W^r \geq 0 \tag{6.6}$$

$$and\ the\ elements\ of\quad P\quad are\ integers \tag{6.7}$$

where $A, B, C, E, F, P, W, W^s, W^r$ in Eqs. (6.2)-(6.7) are as follows:

$$A = \begin{bmatrix} T_{c1} & -T_{c2} & 0 & .. & 0 \\ 0 & T_{c2} & -T_{c3} & .. & 0 \\ . & . & . & .. & . \\ . & . & . & .. & . \\ 0 & . & . & T_{c(n-1)} & -T_{cn} \\ 1 & 1 & 1 & 1 & 1 \end{bmatrix}_{n \times n}$$

$$B = \begin{bmatrix} T_{con(2)}^{\Delta w_0} - T_{con(1)}^{\Delta w_0} + T_{init(2)} + T_{trf(2)}^{\Delta w} - T_{init(1)} - T_{trf(1)}^{\Delta w} \\ T_{con(3)}^{\Delta w_0} - T_{con(2)}^{\Delta w_0} + T_{init(3)} + T_{trf(3)}^{\Delta w} - 2(T_{init(2)} + T_{trf(2)}^{\Delta w}) - T_{con(2)}^{\Delta w} \\ . \\ . \\ \left\{ \begin{array}{c} T_{con(n)}^{\Delta w_0} - T_{con(n-1)}^{\Delta w_0} + T_{init(n)} + T_{trf(n)}^{\Delta w} - 2(T_{init(n-1)} \\ + T_{trf(n-1)}^{\Delta w}) - T_{con(n-1)}^{\Delta w} \end{array} \right\} \\ P_{tot} \end{bmatrix}_{n \times 1}$$

$$W = \begin{bmatrix} (w_1^s - w_2^r) \\ (w_2^s + w_2^r - w_3^r) \\ (w_3^s + w_3^r - w_4^r) \\ . \\ . \\ w_{n-1}^s + w_{n-1}^r - w_n^r \\ 0 \end{bmatrix}_{n \times 1}$$

$$C = \begin{bmatrix} 0 & .. & . & T_{cn} \end{bmatrix}_{1 \times n}$$

$$F = \begin{bmatrix} M_1 \\ . \\ . \\ M_n \end{bmatrix}_{n \times 1} \qquad P = \begin{bmatrix} P_1 \\ . \\ . \\ P_n \end{bmatrix}_{n \times 1} \qquad W^s = \begin{bmatrix} w_1^s \\ . \\ w_{(n-1)}^s \\ 0 \end{bmatrix}_{n \times 1}$$

$$W^r = \begin{bmatrix} 0 \\ w_2^r \\ . \\ w_{(n)}^r \end{bmatrix}_{n \times 1} \qquad E = \begin{bmatrix} r(2N_w + 5N_{tot} + P_1(N_1 + N_l)) \\ r(2N_w + 5N_{tot} + P_2(N_1 + N_l)) \\ . \\ . \\ r(3N_w + 6N_{tot} + P_n(N_1 + N_l)) \end{bmatrix}_{n \times n}$$

Minimization of Eq. (6.2) can be done using such method as the Branch and Bound algorithm [55]. In the examples presented in the next section the mixed integer minimization was performed using the NAG[1] Library [56] routine H02BBF which uses the Branch and Bound method.

---

[1]Numerical Algorithms Group.

6.1 Experimental Results for Benchmark Problems

In this section we use the mixed integer optimization method and obtain the optimal allocation of the training patterns for the T805-20, T805-25 heterogeneous processor networks for two benchmark problems. The time for an epoch corresponding to the optimal pattern distribution is then experimentally obtained. The first benchmark problem is the encoder problem described earlier in Chapter 2.

Our experiment involves mapping a 256-8-256 encoder network with 256 input patterns and 4360 connections including biases on to a heterogeneous processor network with nine T805-20 and seven T805-25 transputers[2]. The external memory for T805-20 is 1 Mbyte and that for the T805-25 is 2 Mbytes. The transputers are connected in a pipelined ring fashion with the first nine transputers in the pipeline as T805-20 model and the last seven as T805-25 model.

Using the mixed integer optimization optimal distribution of patterns is found as 12, 12, 13, 13, 13, 14, 14, 14, 14, 19, 19, 19, 20, 20, 20, 20 for processors 1 to 16 respectively. Allocating patterns in this way the time for an epoch is found experimentally from the heterogeneous network as 2.529 sec. This should be the minimum time for an epoch for this problem on this processor network. Any other way of allocation of patterns should result in higher epoch time.

---

[2]These being the current inventory for transputers in our laboratory.

The second benchmark problem is the sonar problem also described in Chapter 2.

Our experiment involves mapping a 60-24-2 sonar network with 192 input patterns and 1514 connections including biases on to a heterogeneous processor network with nine T805-20 and three T805-25 transputers in a pipelined ring. Again the first nine transputers in the pipeline are of T805-20 model and the last three are of T805-25 model. Using the mixed integer optimization optimal distribution of patterns is found to be 13, 13, 14, 14, 14, 15, 15, 16, 16, 20, 21, 21 for processors 1 to 12 respectively. Allocating patterns in this way the time for an epoch is found experimentally from the heterogeneous processor network as 0.6042 sec. As in the encoder example this should be the optimal (minimum) epoch time for the sonar problem on this processor network.

6.2   Statistical Verification of the Optimal Epoch Time

In order to verify the optimality of the epoch time obtained from pattern allocations resulting from the mixed integer programming, a Monte Carlo simulation study was carried out. Monte Carlo studies are very useful for such analyzes and have been used in the past in relation to learning and training [58]. According to the Monte Carlo method used in the paper, large number of pattern allocations were randomly generated from a uniform distribution for the heterogeneous network (for the benchmark problems) and the epoch times corresponding to each of the random allocation was then experimentally obtained. The sample sizes in the study was varied from 100 to 800 and for each sample size the mean and standard deviation for the epoch time was obtained and compared with the optimal epoch time.

For the encoder problem the Monte Carlo simulations were performed for sample sizes of 100, 200, 500 and 800 for the heterogeneous network. Since the total number of patterns for this problem is 256, a truly random pattern allocation should consider the uniform distribution to be in the range 0 to 256 for each processor. However when this was done we encountered many unrealistic allocations that had zero or very low number of patterns for some of the processors causing the mean epoch time to become very large. In order to have realistic pattern allocations for the processors we chose the mean for the uniform distribution as 16 which corresponds to equal allocation of the 256 patterns amongst the 16 processors. A window of ± 8 patterns from the mean 16 was chosen as the range for the random number generator so as to obtain pattern distributions that are realistic. Further, the random allocation is carried out in such a way that the total number of pattern for a sample run would always equal 256.

The random pattern allocations for each sample were thus generated from a uniform distribution in the range 16 ± 8. The epoch time corresponding to each random pattern allocation was then experimentally obtained. Fig. (6.2) shows the distribution of the epoch time for the encoder problem when the sample size is 500. Fig. (6.3) shows the *mean* $\pm 3\sigma$ for the epoch time for various sample sizes. The optimal epoch time (corresponding to pattern allocations resulting from the mixed integer programming) for the encoder problem is 2.529 sec. As can be seen from Fig. (6.3) the optimal epoch time is always more than three standard deviations (3σ) lower than the sample means.

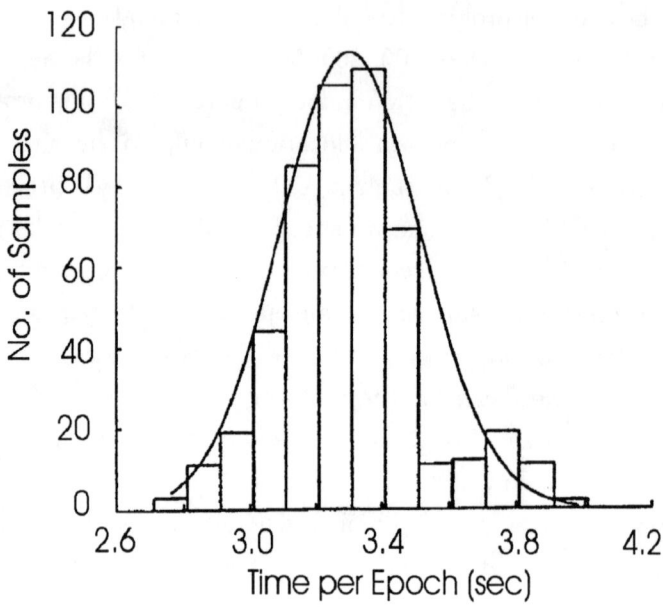

Figure 6.2: Distribution of the time for an epoch for the encoder problem when the sample size is 500. Mean and standard deviation(σ) for the distribution are 3.30 sec and 0.210 sec respectively.

Monte Carlo simulations for the sonar problem were also performed for sample sizes of 100, 200, 500, and 800. The random pattern allocations for each sample were also generated in this case from a uniform distribution in the range 16 ± 8. The epoch time

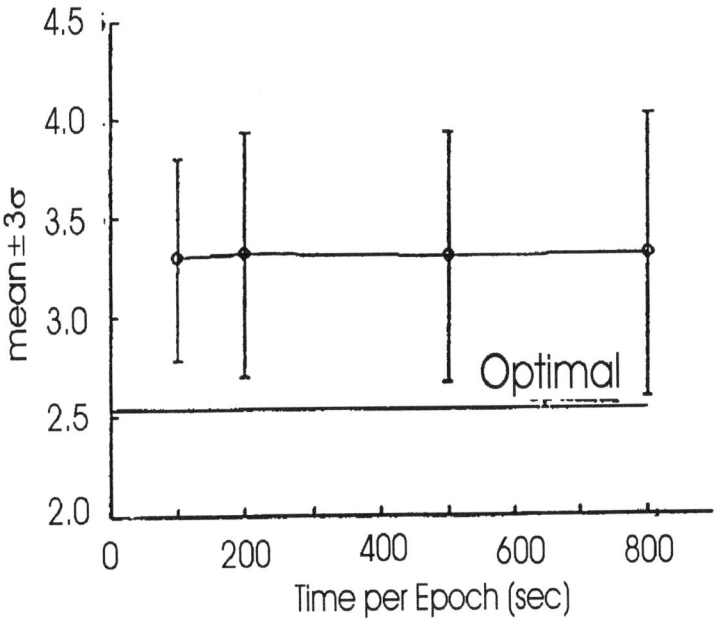

Figure 6.3: Mean$\pm 3\sigma$ for various sample sizes for the encoder problem. The horizontal firm line shows the optimal epoch time.

corresponding to each random pattern allocation was then experimentally obtained as in the previous case. Fig. (6.4) shows the distribution of the epoch time for the sonar problem when the sample size is 500. Fig. (6.5) shows the mean$\pm 3\sigma$ for the epoch time for various sample sizes. The optimal epoch time for this problem is 0.6042 sec. As in the case with the encoder problem the optimal epoch time is more than 3σ away from the sample mean values. The Monte Carlo study thus confirms that the optimal pattern allocations are those obtained from the mixed integer optimization and the resulting epoch time will be the minimum epoch time achievable.

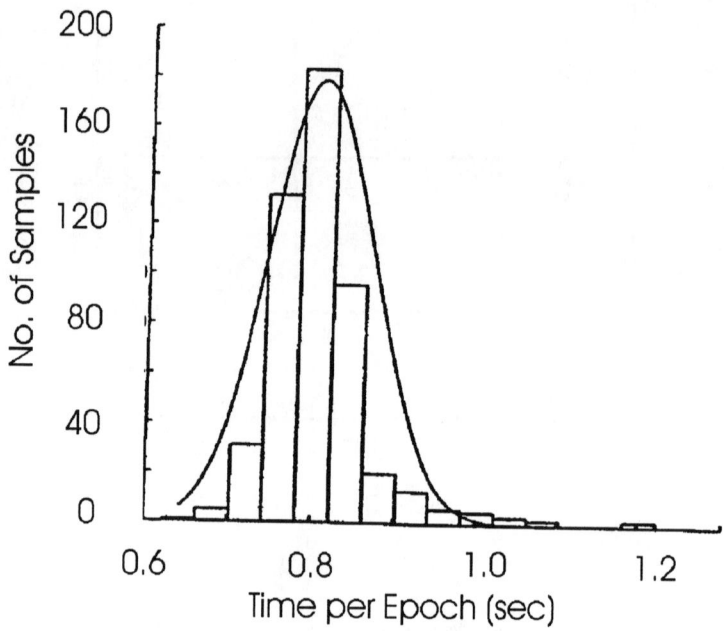

Figure 6.4: Distribution of time for an epoch for the sonar problem when the sample size is 500. Mean and standard deviation(σ) for the distribution are 0.807 sec and 0.062 sec respectively.

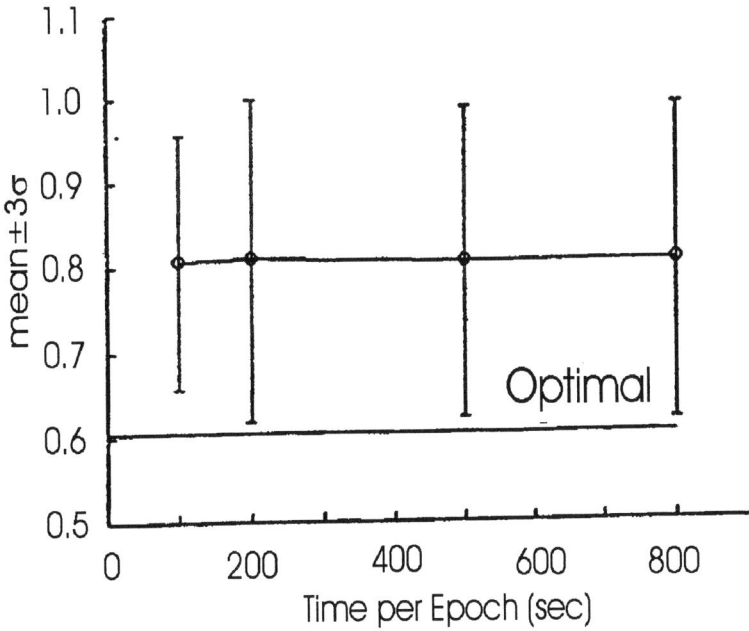

Figure 6.5: Mean$\pm 3\sigma$ for various sample sizes for the sonar problem. The horizontal firm line shows the optimal epoch time.

6.3 Discussion

6.3.1 Worthiness of finding the optimal solution

Although the results in Section 6.2 verify the optimality of the pattern allocation obtained from the mixed integer programming, they do not provide any information about the time taken to find the optimal solution. Since the whole objective of finding the optimal pattern allocation is to obtain a lower training time than any other non-optimal allocations, it is important to investigate whether the time taken to find the optimal allocation itself might offset any reduction in training time accruing from the use of the optimal pattern allocation.

Hence the training time for the optimal allocation was compared with that of non-optimal allocations where the training time for optimal allocation included the time taken to find the optimal solution. Table (6.1) shows the total training time taken by the optimal and three non-optimal allocations for the encoder benchmark problem. The initial weights, learning and momentum rates were all same for all the three cases. The details of the three non-optimal allocations are as follows.

Non-opt-1 : In this case equal number of patterns are allocated to all the processors i.e. 16 patterns for each processor.

Non-opt-2 : In this case unequal distribution of patterns is used for the fast and slow processors. The patterns allocated were 10, 10, 11, 11, 12, 12, 13, 13, 14, 14, 20, 20, 23, 23, 25, 25 for processors 1 to 16 respectively.

Non-opt-3 : In this case the ratio of the number of patterns allocated to a slow processor to that allocated to a fast processor is kept close to the ratio of their speeds. The leftover patterns after such an allocation is distributed among the fast processors. The first nine processors are allocated 14 patterns each, the next three are allocated 18 patterns and the last four are allocated 19 patterns.

Table 6.1: Total time for training for the optimal and two non-optimal pattern distributions for the encoder problem. Convergence was reached after 15,200 epochs.

Allocation Technique	Time taken to Obtain the Allocation (min:secs)	Time taken for training (min:secs)	Total Time taken (min:secs)
Non-opt-1	0:00	729:25	729:25
Non-opt-2	0:00	736:30	736:30
Non-opt-3	0:00	727:40	727:40
Optimal	2:33	640:18	642:51

It can be seen from Table (6.1) that finding the optimal allocation is a worthwhile exercise as the reduction realized in training time more than compensates the extra time needed to find the optimal solution.

Table (6.2) shows the total training time taken by the optimal and three non-optimal allocations for the sonar benchmark problem. The three non-optimal distributions in this case are as follows.

Non-opt-1 :As in the case of encoder in this case equal number of patterns are allocated to all the processors. i.e. 16 patterns for each processor.

Non-opt-2 :In this case an unequal distribution of patterns is used for the slow and fast processors. The patterns allocated are 14,14,14 15, 15, 15, 16, 16, 16, 19, 19, 19 for processors 1 to 12 respectively.

Table 6.2: Total time for training for the optimal and two non-optimal pattern distributions for the sonar problem. Convergence was reached after 300 epochs.

Allocation Technique	Time taken to Obtain the Allocation (Sec)	Time taken for training (Sec)	Total Time taken (Sec)
Non-opt-1	0.0	219.15	219.15
Non-opt-2	0.0	202.31	202.31
Non-opt-3	0.0	218.10	218.10
Optimal	7.29	181.20	188.49

Non-opt-3 : As in the case of encoder the ratio of the patterns allocated to the slow and fast processors is kept as close to the ratio of the speeds of the processors. The first nine processors are allocated 15 patterns each and the last three processors are allocated 19 patterns each.

As in the case with the encoder benchmark here also it can be concluded that finding the optimal allocation is a worthwhile exercise. We feel that for larger problems with tens of thousands of training patterns the reduction in training time due to the use of the optimal pattern allocation will be more pronounced.

Chapter 7

Pattern Allocation Schemes Using Genetic Algorithm

In earlier chapters, it has been shown that finding the optimal distribution of patterns to minimize the time for a training epoch for a heterogeneous network leads to a mixed integer optimization problem. Although solving the mixed integer optimization problem will give the optimal pattern distribution, the computation time required to obtain the solution is usually large. In this chapter, a suboptimal scheme to reduce this computation time is discussed along with how it leads naturally to the use of genetic algorithm for optimization.

7.1 Optimization Algorithm and Computational Complexity

In training-set parallelism the training patterns are distributed amongst the processors and the weights are updated after every epoch i.e. batch learning. The processor network consists of a heterogeneous array of processors (i.e. processors of different speed and memory) connected in a pipelined ring topology. It has been shown in Chapter 6 that finding the optimal distribution of training patterns for such a heterogeneous network to minimize the time per epoch leads to a mixed integer programming (MIP) problem. Although solving the MIP problem gives the optimal distribution, the computation time required to obtain the solution is usually large. This has motivated us to look for sub-optimal solutions to the pattern allocation problem that are computationally fast and produce epoch times that are close to the optimal epoch times. One such sub-optimal scheme is discussed in the following sections. This scheme uses equal allocation if it is possible and if equal allocation is not possible, it uses a heuristic allocation method which modifies equal allocation in such a way that allows faster processors to be allocated more patterns.

7.2   Solution Time for Optimal Pattern Distribution

In this section, we investigate in detail the search time required to find the *optimal* pattern allocation for the NETTALK text to phoneme problem. The optimal solution is obtained by formulating and solving the related MIP problem described in Chapter 6. The neural network used for this problem has 203 input units, 60 hidden units, 26 output units and 1500 training patterns. The MIP for the NETTALK was solved on a VAX 9000 machine using the NAG library routine H02BBF [56] which uses the branch and bound search method.

A simulated heterogeneous processor array of fifty processors comprising a mixture of forty T805-20 and ten T805-25 transputers connected in a pipelined ring fashion is used as the 'hardware' network. The first forty processors in the pipelined ring are T805-20 and the last ten processors are T805-25.

The solution time for the MIP depends on the number of search nodes used in the branch and bound method. Larger the number of search nodes, more is the time taken to find the solution whereas small number of search nodes would take less time but may give non-optimal solution. The search nodes for the NETTALK was varied from 4,000 to 40,000 and the time for finding the solution was recorded for each case. The first mixed integer solution is only obtained after 3632 search nodes. The VAX CPU time to obtain the pattern distribution is 12 minutes and the corresponding time per epoch is found as 13.054 sec. Table (7.1) shows the time taken by the NAG routine to solve the MIP problem for search nodes of 4,000, 6,000, 8,000, 10,000, 20,000 and 40,000. Fig. (7.1) shows the time for an epoch for the corresponding pattern distribution on the simulated hardware. As can be seen from the figure the time per epoch shows little improvement beyond 8,000 search nodes. However the time taken to solve the MIP increases considerably with increase in the search nodes as shown in Table (7.1).

Also shown in Fig. (7.1) is the time for an epoch resulting from a distribution obtained from a Simplex solution to the pattern allocation problem. In the Simplex [59] solution all integer constraints are relaxed and the optimal pattern distribution is obtained by solving the resulting linear programming problem. The pattern distribution resulting from the Simplex solution will not be realistic as it will comprise fractional patterns in the allocation. However the epoch time corresponding to these non realistic allo-

Table 7.1: Time taken by the NAG routine to solve the mixed integer problem

Number of Search Nodes	4,000	6,000	8,000	10,000	20,000	40,000
CPU Time Taken to solve the MIP problem (hr:min)	0:13	0:19	0:24	0:29	0:57	1:51
Time per Epoch (sec)	12.931	12.931	12.909	12.909	12.909	12.909

cation can be calculated and used as an unattainable lower bound in the comparison of optimal and sub-optimal methods.

7.3 Sub-optimal Method: Heuristic Distribution

It can be seen from the Table (7.1) that even for the 8,000 search node, the solution to the MIP took 24 minutes of CPU time on a VAX 9000 machine. This motivated us to look for a heuristic sub-optimal allocation method that is computationally fast and produces performance comparable to optimal solution.

The heuristic method uses a pattern distribution scheme in which equal division of patterns amongst the processors is used

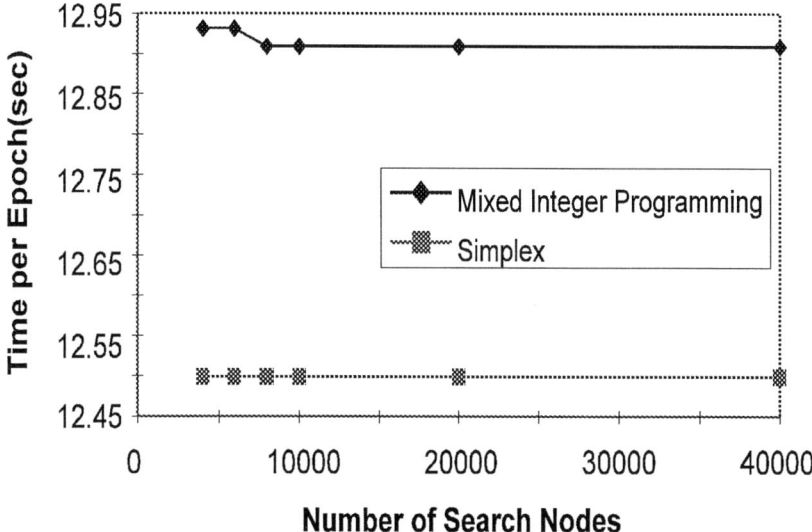

Figure 7.1: Time per epoch versus number of search nodes for the NETTALK problem

if the number patterns in the training set is an integer multiple of the number of processors. If equal division is not possible then the following strategy is used. Allocate $\lfloor P_{tot}/n \rfloor$ patterns to the n processors. Allocate the remaining P_{tot} mod n patterns starting from the fastest processor. Mathematically the number of patterns allocated to the ith processor is given by

$$P_i = \lfloor P_{tot}/n \rfloor + R$$

where R is a binary number.

$$R = 1 \; if \; (n - i) < (P_{tot} \; Mod \; n)$$

$$R \;=\; 0\,if\,(n-i) \geq (P_{tot}\,Mod\,n)$$

This heuristic allocation scheme was used for pattern allocation for the NETTALK problem and the results are presented in Table (7.2). The time for an epoch is obtained for the same simulated heterogeneous processor network of forty T805-20 and ten T805-25 transputers. Table (7.2) shows the time per epoch as well as the time taken to obtain the pattern allocation for this sub-optimal scheme. The time for an epoch from the Simplex solution and the optimal mixed integer (8,000 nodes) solution is also given in the table to show how well this sub-optimal method performs relative to the MIP and Simplex solutions.

Table 7.2: Timing comparison for different pattern allocation schemes

Allocation Scheme	Time per Epoch (sec)	CPU Time Taken to obtain the allocation (min:sec)	% Diff. wrt Simplex Distribution for Time per Epoch
Heuristic Distribution	15.764	0:3	26.12
Optimal(8000 nodes) MIP solution	12.909	23:44	3.28
Simplex Distribution	12.499	0:3	0

The heuristic method, although very fast in obtaining a sub-optimal solution, produced an epoch time which is about 25% more than that of the MIP optimal solution. This motivated us to look at genetic algorithm (G.A.) as a better sub-optimal method of solution.

7.4 Genetic Algorithm For Pattern Allocation

The main objective of finding the optimal pattern allocation is to minimize the total training time and this objective will be defeated if the method used to obtain the optimal pattern allocation itself takes large amount of computing time. Finding the optimal pattern allocation using the MIP method does take considerable amount of computing time. In the last section, we studied a suboptimal method to reduce this time but the problem with that method was that it was ad-hoc and the resulting solution was also not close to optimal solution. Hence, to approach this problem in systematic way we were motivated to look for Genetic Algorithms(GA) based allocation schemes that are computationally fast and could produce epoch times which are closer to MIP epoch time.

The genetic algorithm used in this study is a modified version of the simple genetic algorithm [60, 61]. Modifications are made so as to meet the needs for the pattern distribution problem described earlier. The basic steps for the genetic algorithm are as follows:

1. Obtain an initial population.

2. Obtain the fitness of each chromosomes in the population.

3. Select chromosomes to form the intermediate population.

4. Perform the crossover operation on the population.

5. Perform mutation operation on the population.

6. Test if the stopping criteria is met.

7. If stopping criteria is met end the search, if not, go to step 2.

The origins of genetic algorithm is based on the principle of the survival of the fittest among a given population. Given an initial population, there are members that are fitter than others. In a competition, it is expected that fitter members will survive while the weaker ones will die off. These surviving members will then recombine to give offsprings. These offsprings are expected to have the characteristics of their natural parents. Moreover, these offsprings are supposed to replace their parents in the next generation. As time goes by, better and better offsprings will be produced. This will eventually lead to the formation of a population that is made up of all the elite members. When this happens, no more better offsprings can be produced and the genetic algorithm terminates.

The details of steps 1-6 are discussed in the following sections.

7.4.1   Obtaining the Initial Population

Before the initial population can be obtained, it is necessary to have a suitable representation for each chromosome in the population. In our parallel implementation of the backpropagation algorithm, the transputers are connected in a pipelined ring topology with each transputer in the pipelined ring allocated a certain number of training patterns. In our genetic representation, the length of each chromosome is n, where n is the number of processors in the pipelined ring. The ith gene, $1 \leq i \leq n$, will correspond to the ith processor in the pipelined ring and its allele will hold the number of patterns distributed to that processor. Each allele will be represented by a floating number instead of the usual binary number. This floating number representation will simplify the recombination operations defined later in Section 7.4.4.

The initial distribution of the patterns in each chromosome is obtained by allocating a certain number of training patterns to each gene. Assuming P_{tot} is the total number of training patterns in the training set, each gene will be allocated $\mu + \sigma$ number of patterns where $\mu = \lfloor P_{TOT}/n \rfloor$ and σ is a uniform random number in the range $\lfloor \pm \mu/2 \rfloor$. In this random allocation, there is a very high chance that the sum of all the patterns allocated in a chromosome may not be equal to the total number of training patterns in the training set. As this is a necessary constraint to be satisfied, a chromosome obtained in this way has to be rejected. In order to ensure that the sum of all the patterns is equal to the total number of training patterns, we perform a normalization for each chromosome. This is done by summing up all the patterns present in the chromosome and dividing each gene value by this number. The result is then multiplied with the total number of training patterns, P_{tot}. If at the end of all these a non integer number results, then it will be truncated to the nearest integer. This truncation produces a side effect in that some patterns will still be left unallocated. These will be allocated to the genes which had the highest remainder before truncation. If there are two or more genes that had the same remainder, then the rightmost gene will be allocated with the extra pattern. To illustrate, consider the following example where there are 100 patterns to be allocated among 6 processors. This is illustrated in Fig. (7.2). In this case, μ will be $\lfloor 100/6 \rfloor = 16$ and $\sigma = 16/2 = 8$. Assume that the random distribution of patterns (g_i) in a particular chromosome is as follows: 14, 15, 12, 20, 17 and 12. Then, the total number of patterns in the chromosome will be $\sum_{i=1}^{i=n} g_i = 90$. Since the sum of all the gene values is not equal to the total number of training patterns (i.e. 100), the chromosome has to be normalized. This is done by truncating the gene value of each gene after it has been divided by 90 and multiplied by 100. For example, for gene g_1, the normalized gene value (g_i') becomes $\lfloor 14 * (100/90) \rfloor = 15$. Continue in this way, the gene values, g_i's for all the genes after truncation will be 15, 16, 13, 22, 18 and 13. The sum of these modified gene values is $\sum_{i=1}^{i=n} g_i' = 97$. Since $P_{tot} = 100$, there are three patterns still to be allocated. These three patterns will be allocated to those gene with the highest remainder before truncation. If there is a tie among the remainders, the gene located furthest away from the root processor in the pipelined ring will get the allocation. The resulting pattern distribution for this example is 16, 17, 13, 22, 19 and 13. By this normalization, it is ensured that the sum of all the patterns in the chromosomes will always be equal to the total number of training patterns (P_{tot}). Such a normalization will also ensure that the fitness of each chromosome will be reasonably high compared to an arbitrary allocation of patterns in each chromosome.

Total Number of Patterns = 100

Before Normalization

| 14 | 15 | 12 | 20 | 17 | 12 | Sum = 90 |

Before Multiplying By 100/90

| 15.6 | 16.7 | 13.3 | 22.2 | 18.9 | 13.3 | Sum = 100 |

After Truncating

| 15 | 16 | 13 | 22 | 18 | 13 | Sum = 97 |

After Normalization

| 16 | 17 | 13 | 22 | 19 | 13 | Sum = 100 |

Figure 7.2: Figure showing the process of normalization

7.4.2 Calculating the Fitness of Each Chromosome

The fitness for each chromosome is defined as

$$Fitness = \frac{T_{epoch}^{serial} - T_{epoch}^{parallel}}{T_{epoch}^{serial}} \qquad (7.1)$$

where T_{epoch}^{serial} is the time per epoch if the pattern allocation problem is to be run on one processor and $T_{epoch}^{parallel}$ is the time per epoch taken if the problem is to be run on n processors. The above equation will scale the fitness of each chromosome to be between 0 and 1.

7.4.3   Selecting Chromosomes for the Intermediate Population

After obtaining the initial population, it is necessary to select from this population an intermediate population. It is this intermediate population that will undergo recombination in order to obtain the next population or members for the next generation. For this study, the intermediate population is selected using the tournament selection [62] scheme. In a preliminary study [63], we compared the tournament scheme with the roulette wheel selection [64] scheme and found the tournament selection scheme to perform better in most cases. This may be because for our problem, there is close competition among the chromosomes for selection. Due to this close competition, the roulette wheel is unable to select efficiently the fit chromosomes for the intermediate population. This results in having many weaker chromosomes being selected in the intermediate population. However, in tournament selection, the best chromosomes are definitely selected. This will result in a fitter intermediate population. As the number of generations increases, the population obtained will thus be better. However, the use of tournament selection does come with a certain trade off. For this scheme, more computation time is needed as there is a need to pick the best chromosome among a specified number of chromosomes.

7.4.4 Crossover Operations

In the simple genetic algorithm, after obtaining an intermediate population, the next step is to perform recombination and obtain the next generation. Recombination is done via crossover and mutation. The crossover operations are performed with a probability of p_c on two randomly chosen chromosomes from the intermediate population. That is, each pair of chromosomes has a chance of p_c that they undergo crossover. If they are not required to be recombined, this pair of chromosome will be duplicated in the next population. The following sections discuss each of these crossover operations in detail and their effects to the overall fitness of the next population will be studied later.

A listing of the crossover operators discussed in [63] are:

- One Point Crossover (XOVER1)

  The one point crossover operator operates on two chromosomes chosen at random from the intermediate population. A crossover point is then chosen at random. This crossover point is any random integer from 0 to $n - 1$ where n is the length of a chromosome or the number of transputers in the pipelined ring. Once a crossover point is chosen, the genes after the crossover point will be interchanged from the two parent chromosomes to obtain two offsprings. The two children chromosomes obtained in this way will not necessarily have a pattern distribution which has a number of patterns equal to the total number of training patterns in the neural network. Normalization as described in Section 7.4.1 is then performed to ensure that the total value of all the genes tally with the sum of all the training patterns.

- Two Point Crossover (XOVER2)

  For this crossover operator, instead of selecting one position
  for crossover, two positions are chosen at random. The genes
  in between these two positions are interchanged from the parent chromosomes to obtain the two offsprings. The whole of
  each of the two offsprings are then normalized.

- Two Point Part Crossover (XOVER3)

  This crossover operator is similar to two point crossover except that the normalization is performed only for those genes
  between the two crossover points.

- Uniform Crossover (XOVER4)

  For uniform crossover, a mask of length L is chosen. This
  mask is made up of a random combination of zeros and ones.
  The first parent chromosome will only interchange its gene
  value whenever a zero is encountered in the mask; otherwise,
  its original gene value will be duplicated in the offspring. The
  second parent chromosome will exchange its gene value with
  the first parent chromosome only when a one is encountered
  in the mask; otherwise, it will retain its gene value in the
  offspring. After this recombination, each of the two offsprings
  will then undergo normalization.

Out of the crossover operators discussed above, it was found
that, in general, the performance of the Uniform Crossover was the
worst. This can be attributed to the fact that it is very much like
a mutation operator as a lot of exchanges takes place. There are

too many disruptions occurring in the search causing the search to 'look' random. The One Point crossover on the other hand performs better. However, because it causes a disruption only at one location, not many new chromosomes are found. That is, not many good hyperplanes are produced. The results of using this are also not very favorable. The performance for the Two Point Crossover and the Two Point Part Crossover operators are the best. In these, there are more disruptions to allow for a good coverage of the solution space and at the same time allow for more exploration. Out of these two operators, the Two Point Part Crossover performs better in terms of the number of generations needed for convergence. Two Point Crossover requires many more generations than Two Point Part Crossover to achieve the same performance. The Two Point Part Crossover operator has a localized effect as only genes between any two crossover points are normalized. Since the swapping of two genes which are close together has a lesser impact on the fitness value than swapping the positions of two genes far apart, it is expected that the time per epoch for Two Point Part Crossover will have lesser change. Since not much improvement in the epoch time is obtained by Two Point Crossover over the Two Point Part Crossover, in all further discussion in this study, only the Two Point Part Crossover operator will be considered.

7.4.5   Mutating the Chromosomes

Mutation is another important recombination operation that has to be done after the crossover operation. In this study, the Uniform Mutation scheme is chosen. In this mutation scheme, when a particular chromosome is required to undergo mutation with probability p_m, each gene value g_i will take on the value $g_i \pm \sigma$ where σ is a random integer in the range 0 and $\lfloor \mu/2 \rfloor$; μ is the average number of training patterns. After each gene has been mutated, the whole chromosome will then be normalized so that the sum of the gene values will be equal to the total number of training patterns. Fig. (7.3) shows a chromosome before and after uniform mutation is applied.

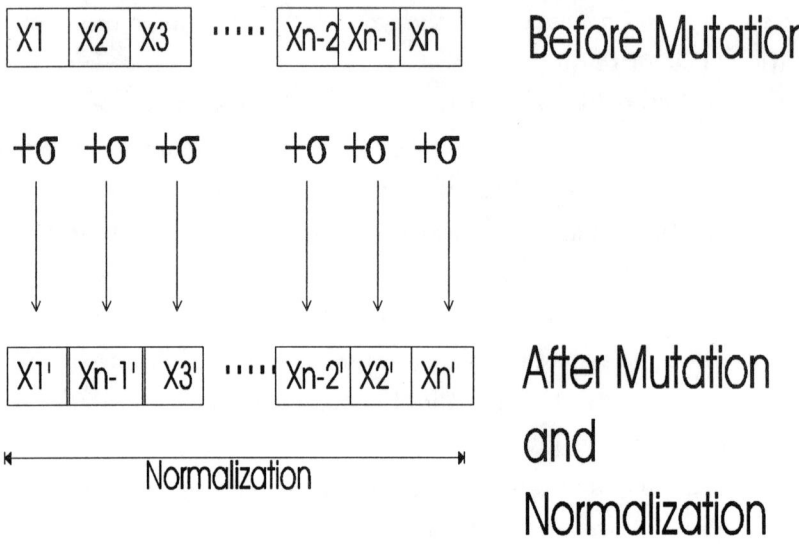

Figure 7.3: Figure illustrating the uniform mutation operator

7.4.6   Stopping Criterion

The preliminary stopping criterion used in this study was to search through a total of 10,000 generations and then stop. The generation at which the best epoch time occurs is recorded for comparison purposes. This method of comparison sheds some light on how good a particular type of recombination operator is with respect to another. A more sophisticated stopping criterion is discussed in Section 7.6.2.

7.5 Comparison between Genetic Algorithm and MIP

Preliminary set of experiments are carried out to compare the performance of the pattern allocations produced by genetic algorithm to those produced by mixed integer programming [51, 65, 66]. The experiments conducted were aimed at finding the effects of different crossover operators used and also find out the effects of complexity of the problem. For the G.A., the population size chosen is 150 and the tournament chosen is 50. The generation limit for the encoder problem is set as 5,000 and the generation limit for the NETTALK problem is set as 10,000. No elitism is used in the simulation because it was found that there is not much effect to the final results [63]. The crossover probability used is 0.75 while the mutation probability used is fixed at 0.1.

Tables (7.3) and (7.4) show the results obtained for the 256-8-256 encoder problem and for the NETTALK problem respectively. In each of these tables, 'CPU Time Taken' for G.A. refers to the average VAX CPU time needed to obtain this solution out of five runs. Likewise, the 'Epoch Time' for G.A. is the average time per epoch resulting from the five runs. Also shown in these two tables are the results obtained using MIP as well as that obtained using an equal pattern allocation scheme.

Table 7.3: Table showing the comparison of solutions between genetic algorithms and mixed integer programming for the encoder problem

Technique Used	256-8-256 Encoder	
	Epoch Time (msec)	CPU Time Taken (min:sec)
Genetic Algorithm	2324	0:04
Mixed Integer Program-ming	2324	3:40
Equal Distribution	2690	0:0

It can be observed from both Table (7.3) and Table (7.4) that when equal pattern allocation scheme is used, the time per epoch obtained is the highest compared to those obtained from the MIP and G.A. It should be noted that although the difference in the time for an epoch between equal and the other two solutions is only around 350 msec for the encoder, the savings in total training time will be significant since the number of training epochs needed for convergence is usually large. This is usually true for most problems and hence indicates that it is worthwhile to search for optimal solution. Also, from Table (7.3), for the 256-8-256 encoder problem, the genetic algorithms found the optimal solution in 4 sec whereas the MIP solution took 3 min 40 sec. This can be attributed to the high disruptive nature of the mutation operator which tends to

Table 7.4: Table showing the comparison of solutions between genetic algorithms and mixed integer programming for the NETTALK problem

Technique Used	NETTALK	
	Epoch Time (sec)	CPU Time Taken (min:sec)
Genetic Algorithm	13.06	14:59
Mixed Integer Programming	12.909	24:0
Equal Distribution	15.76	0:0

create more opportunity for chromosomes to explore more regions. Thus convergence tends to speedup in this case.

In the MIP solution for the NETTALK problem, the search for the computation of the optimal distribution was terminated after 8,000 search nodes as no significant improvement was observed for more search nodes. From Table (7.4), it can be seen that the solution time taken is 24 min and the time per epoch is 12.909 sec. The G.A. could not obtain an epoch time as low as 12.909 sec even after 10,000 generations.

This simulation shows that the basic recombination operations and selection scheme alone are capable of outperforming the mixed integer programming solution for smaller problems. But when the problem complexity increases, the basic genetic algorithm produces a performance roughly in the same order of magnitude as the mixed integer programming.

7.6 Inclusion of *'A Priori'* Information

A way to improve the performance of the genetic algorithm is to reduce the tendency to search the entire population space. This is best done if we can combine some *'a priori'* information peculiar to the problem during the genetic search. For the present problem, due to the synchronization needed among the processors and due to the arrangement of the processors (slower processors preceding the faster processors in the pipelined ring), one can deduce the following. Pattern allocations should follow an ascending order starting from the first processor in the pipelined ring. This is clear if we look at the results in Section 6.1. From Fig. (6.1), since the downstream transputer in the pipeline must wait for the upstream transputer before the next stage of computation can be carried, more patterns should be allocated to fill up the gap due to the waiting time introduced during synchronization. Thus, one *'a priori'* information is to always arrange the patterns (allele values) in an ascending order after each recombination. The re-ordering can be easily implemented using the quick sort algorithm [67].

Table (7.5) shows the performance of the G.A. when *'a priori'* information is added to the basic genetic algorithm for the NETTALK problem with 40 T805-20 and 10 T805-25 transputers. Table (7.6) shows the performance of the G.A. for the same

NETTALK problem without the *'a priori'* information.

Table 7.5: Table showing the effect of adding *'a priori'* information to genetic algorithms

Technique	With *'A Priori'* Information	
	Epoch Time (sec)	CPU Time Taken (min)
G.A. with XOVER3	12.68	16:32
Mixed Integer Programming	12.91	24:00

Table 7.6: Table showing the effect of not adding *'a priori'* information to genetic algorithms

Technique	Without *A Priori* Information	
	Epoch Time (sec)	CPU Time Taken (min)
G.A. with XOVER3	13.06	14.99
Mixed Integer Programming	12.91	24.00

The results from Table (7.5) when compared with those in Table (7.6) show that it is indeed beneficial to incorporate *'a priori'* information into the basic G.A. Moreover, the time per epoch with *'a priori'* information is better than even that of the MIP. This shows that genetic algorithms can outperform the MIP if *'a priori'* information about the solution is incorporated in the search.

7.6.1   Different Stopping Criteria

One problem with the stopping criterion described in Section 7.4.6 is that it is necessary to run through all the 10,000 generations and then look back to find out which recombination method is good. However, there are many cases whereby only a few generations are needed for convergence. Hence, saying that particular recombination takes only few number of generations to converge is not fair because in actual fact, 10,000 generations are used in all cases. Moreover, 10,000 generations may not be necessarily big enough for more complex problems. Hence, there is a need to obtain a better stopping criterion.

There are many different stopping criteria used in the literature for G.As. In this section, we discuss in detail the pros and cons of various commonly employed stopping criteria and indicate why they are unsuitable for our problem.

1. **The number of generations reaches a certain limit. [68, 69]**

   For this stopping criterion, all we have to do is to select a large number of generations and allow the G.A. to run all the way to that generation and then stop the search. However, determining this generation limit is not easy. It depends on the problem complexity. More generations are needed for complicated problems.

2. **The maximum allowable simulation time is reached. [69]**

   If the user is allocated only a certain amount of CPU time

and program memory, one method to stop the simulation is to allow the G.A. to run until the maximum CPU limit is reached. This can be done with the largest population size possible (within memory limits) as the number of generations needed for convergence using a large population size should be low [62]. The problem with this stopping criterion is that we may use up too much CPU time or memory for smaller problems.

3. **The solution is within a certain specified error bound of the desired solution. [68]**

   This criterion is useful for mathematical problems where we know for sure that the error bound is achievable. If not, the simulation could be running forever. For example, if we are trying to find a value of x such that $f(x) = 0$, where $f(x)$ is a 10-order polynomial equation, we can stop the simulation when $f(x)$ is close to zero because this is a possible error bound.

   However, this criterion cannot be applied to our problem because we are unable to know beforehand what is the best time per epoch unless we are given the pattern distribution which is what we want to find out.

4. **The normalized standard deviation of the fitness falls below a certain threshold.**

   When the normalized standard deviation of the fitness is low, it indicates that there is little variation in the fitness of the population. The normalization is done by dividing the standard deviation of a population with the population size. When this happens, the fitnesses of every chromosome is more or less identical. This uniformity indicates that not much improvement in the fitness can be expected in future generations. This form of stopping criterion can be used for multimodal problems where different optimal chromosomes could have resulted in the same optimum fitnesses. However, for our pattern allocation problem, the fitnesses of the chromosomes are quite close even from the start. Moreover, our problem is very sensitive to the mutation operator which makes this criterion inefficient.

5. **The ratio of the current average of the population fitness to the previous average of the population fitness is more or less unchanged. [69, 70]**

   This stopping criterion has the same disadvantages as the previous criterion.

6. **The population is not creating any better offsprings.**

   Offsprings that are produced after the recombination operations should be generally better than their immediate parents. If there comes a time when a high proportion of the parents are found to be better than the offsprings, it means that there is not much improvement and that G.A. is not 'performing' well. This indicates that it is time to stop the search. For our problem the fitness is very sensitive to the pattern allocations, e.g. a swap in two patterns could change the fitness quite a bit. This causes a fair mix of weak and fit chromosomes to exist in the population even after several generations. This makes it difficult to use this stopping criterion.

7. **The standard deviation of each locus in each chromosome falls below a certain threshold. [68, 71]**

This stopping criterion is very useful if the problem nature is unimodal. For a unimodal problem, there is only one optimum solution. Thus, chromosomes that are weak will be rejected eventually leaving behind the fitter ones. Since for a continuous unimodal problem, the fitness of the chromosomes in the neighborhood of the optimum chromosome should be close to each other. This means that the standard deviation of the allele for each gene in a population will be low. When this happens the search is stopped. However, for our problem, which is multimodal, this stopping criterion cannot be used.

7.6.2 Proposed Stopping Criterion

The stopping criteria described in Section 7.6.1 all have some problems or other when applied to our pattern allocation problem. In this section, a new stopping criterion is proposed. It attempts to solve the problem of excessive number of generations and CPU time (cases 1 and 2) by stopping on its own. It also attempts to solve the error bound problem (case 3) by not considering any error function for stopping. This stopping criterion should also be useful for both unimodal and multimodal problems (case 7). It should also be insensitive to parameters variations, for example, population sizes or mutation probabilities (cases 4,5,6).

The stopping criterion proposed here plays the role of finding a compromise solution between the best epoch time and the computation time needed to find this best epoch time. It is based on the assumption that as the G.A. progresses, there will come a time when a large number of generations is needed to bring about only a small improvement in the fitness value. When this happens, which can be deduced from the gradient of a plot of best fitness against number of generations needed to obtain that fitness, a simple criterion is used to decide whether or not to stop the search. The main idea of this procedure is illustrated in Fig. (7.4).

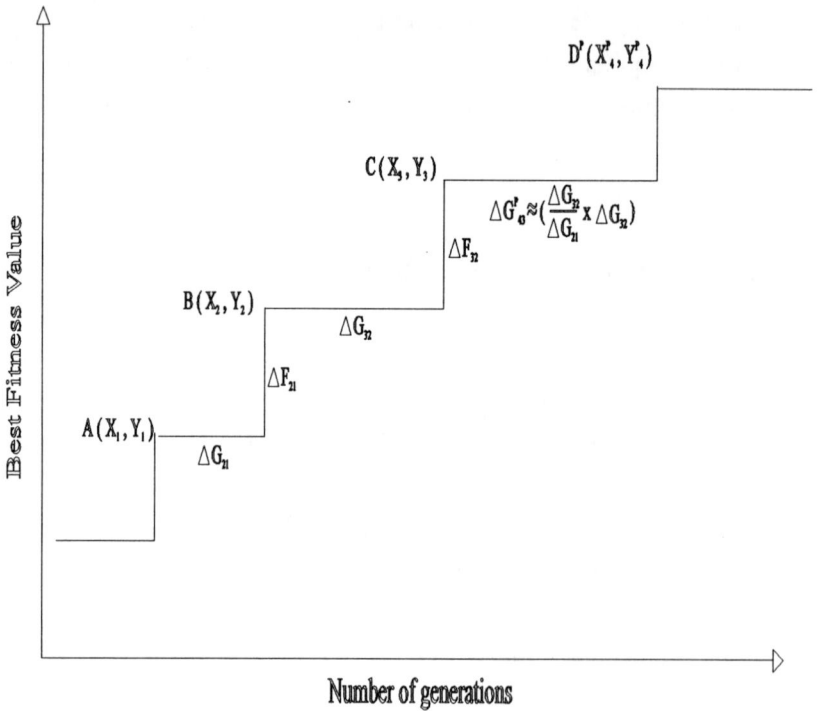

Figure 7.4: Proposed convergence test

In Fig. (7.4), A, B and C are three points on the plot of best fitness against the number of generations. Point D_P is a predicted point where the next best fitness value may occur. If ΔG_{32} is greater than ΔG_{21} and ΔF_{32} is smaller than ΔF_{21}, it indicates that the G.A. is evolving more slowly and is beginning to converge. When this phenomena is detected, the following criterion is used to stop the search. According to this criterion, the improvement in fitness value should occur within a generation gap of $(\Delta G_{32}/\Delta G_{21}) * \Delta G_{32}$. If no improvement occurs within this generation gap, then convergent is assumed to occurred and the search is stopped.

This stopping criterion is only applicable if the G.A. is able to find at least three fitness improvements. If not, the test will never be carried out. In the situation where the population size is extremely large, there is some chance that the optimal solution is found in one or two generations. To tackle this problem, it is assumed at the start of the search that a better fitness value can be found after 500 generations. If this does not happen, the search stops.

As an illustration, assume that we make use of this stopping criterion and allow the G.A. to progress until this criterion is met. Fig. (7.5) shows the epoch time against number of generations needed for the NETTALK problem for a typical run.

For this example, when the stopping criterion described in this section is used, the G.A. obtained its best epoch time of 13.18 sec after 245 generations. This is indicated by the vertical arrow ($\uparrow$) in the figure. However, if we allow the G.A. to run to its maximum generation limit of 10,000, the best epoch time is 13.04 sec occurring after 6773 generations. The reduction in epoch time obtained is about 1.07% whereas the additional search time required to obtain this reduction is 2775%!

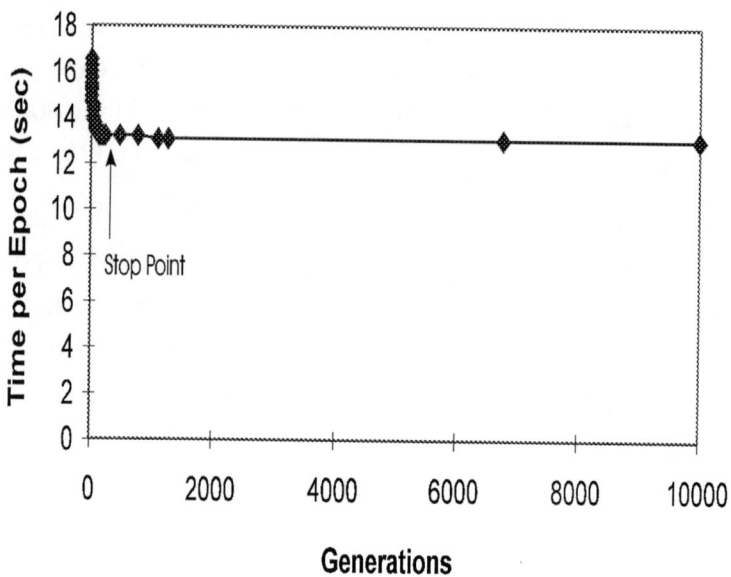

Figure 7.5:   Epoch time versus generations needed for the NETTALK problem

7.7   G.A. with the Proposed Stopping Criterion Versus MIP

In this section, we compare the performance of the proposed stopping criterion with the preliminary stopping criterion described in Section 7.4.6. To make the comparison more meaningful, the effect of including *'a priori'* information is also considered.

Table (7.7) shows the search time and the epoch time for the genetic algorithms for the NETTALK problem. The last row in Table (7.7) shows the corresponding times for the MIP solution.

Table 7.7: Table showing comparison of results between MIP (with 8,000 search nodes) and G.A. with proposed stopping criterion

	Preliminary Stopping Criterion		Proposed Stopping Criterion	
	Epoch Time (msec)	CPU time (min)	Epoch Time (sec)	CPU Time (min)
G.A. Without 'a priori' Info	13.06	14.99	13.16	6.77
G.A. With 'a priori' Info	12.68	16.54	12.70	2.16
Mixed Integer Solution	12.91	24.00	12.91	24.00

It is clear from the table that the proposed stopping criterion takes less time to obtain a solution close to what is achieved by the stopping criterion of Section 7.4.6. The MIP solution is clearly the most extravagant in terms of search time and even in cases where it produces shorter epoch time, the improvement is marginal.

To summarize, this chapter presented the use of genetic algorithms in finding the optimal distribution of training patterns among a heterogeneous array of processors connected in a pipelined ring topology. It is found that when the number of processors used is small, genetic algorithms can obtain the optimal distribution in a very short time. However, when the number of processors used is large, use of 'a priori' information greatly improves the performance of the G.A. The time to find the optimal solution for G.A. is always shorter than the mixed integer programming.

Another aspect is the study of the different stopping techniques used in genetic algorithms. It is shown that many of the commonly employed stopping techniques are not suitable to the pattern allocation problem considered here. A new stopping criterion that serves the role of finding a compromise between the optimal solution and the search time needed to find this optimal solution is proposed. The performance of this stopping criterion is compared with other stopping techniques. It is found that in most cases, the proposed stopping criterion manages to find a good stopping point. Although this stopping criterion is developed with reference to the pattern allocation problem, we feel it can be used as a general stopping criterion for other genetic algorithms applications.

Appendix A

Comparison Between Pipelined Ring Topology and Ring Topology

In this appendix, a comparison between the pipelined ring topology, which is used in our study, and the common ring topology is carried out. The pipelined ring topology used in this study is an extension of the ring topology in that we allow the communications and data transfers to be bi-directional. Another difference is that in the pipelined ring topology, no broadcasting of data is performed. The objective of this appendix is to find out under what conditions the optimal pipelined ring is superior to the optimal ring topology. To make the comparison meaningful, we consider the theoretical optimal time per epoch for both the topologies.

A.1 Theoretical Optimal Epoch Time for Pipelined Ring Topology

In this section, the optimal time per epoch for the pipelined ring topology is given.

The steps involved in executing one epoch of our parallelized backpropagation algorithm when the pipelined ring topology is used is given in Section 3.2 of Chapter 3 and is reproduced below.

1. The weight changes and bias changes for the current epoch are initialized to zero at the start. The time taken for this operation is defined as $T_{con}^{\Delta w_0}$.

2. After initialization, the forward pass and the backward pass of the backpropagation algorithm is performed for each training pattern assigned to the processor. The time needed for this operation is T_c.

3. The total weight change and error due to all the patterns assigned to the transputer are calculated by accumulating the weight change and error calculated for individual patterns in step 2. It takes $T_{con}^{\Delta w}$ amount of time for this operation to complete in each processor.

4. The first pipe processor sends its accumulated weight change and error to its downstream (right) neighbor which adds its own accumulated weight change and errors and sends the updated sums to its right neighbor which in turn updates the sums and sends to its right neighbor and so on. The last pipe processor would thus receive the accumulations of all the processors to its left, adds its own weight change and error and obtains the total weight change and error for the epoch. It then adds the weight change to the existing weight values and obtains new values for all weights in the neural network. The new weight values are then passed upstream by each pipe processor except the first. In an n transputer network, there will be $n - 1$ transfers of weight change and

error and another $n - 1$ transfers of weights. The time taken to send or receive one set of weight change and error from a upstream pipe processor to a downstream pipe processor is $T_{init} + T_{trf}^{\Delta w}$. The time taken to send or receive one set of updated weights back from a downstream pipe processor to a upstream pipe processor is $T_{init} + T_{trf}^{w}$. T_{init} refers to the time taken to initialize or start off a communication process while $T_{trf}^{\Delta w}$ and T_{trf}^{w} represent the time taken to transfer a packet of weight changes or weights across two communication channels respectively.

5. At the same time as the weights are updated, the last processor sends the total error to the root processor which checks whether convergence is reached. The time taken for this operation is small and is neglected.

In Chapter 3, we obtained the time per epoch for the pipelined ring topology for a homogeneous array of processors. We notice that to optimally distribute all the training patterns among a homogeneous array of processor, the best way is to distribute the patterns in such a way as to minimize the waiting times between processors. Practically, all the waiting times can never be made zero as this requires fractional patterns to be allocated to processors which is not feasible. However, from a theoretical point of view, if we set all the waiting times to zeros, the best epoch time will result. If we consider this ideal case, we obtain the timing diagram as shown in Fig. (A.1).

From the diagram, we have between pipe processors 1 and 2,

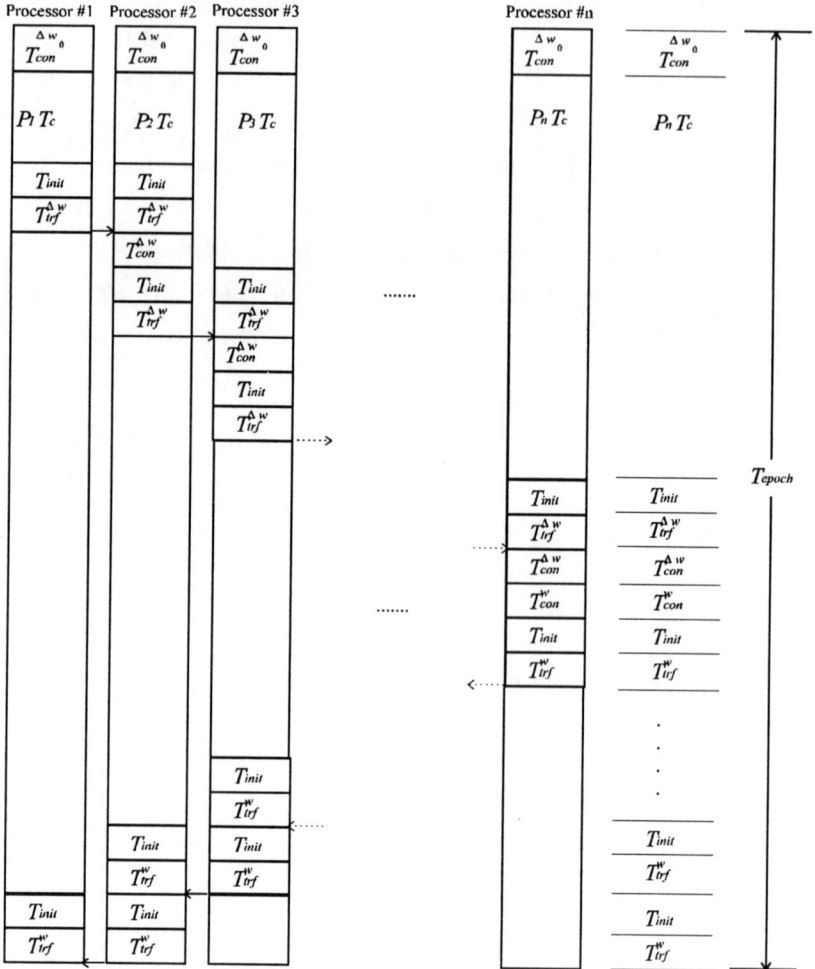

Figure A.1: Timing Diagram for an epoch for a homogeneous array of transputers with theoretical optimum pattern distribution for the pipelined ring topology.

$$P_1 T_c = P_2 T_c. \tag{A.1}$$

Between processors 2 and 3 with no waiting time in processor 3, we have,

$$P_2 T_c + T_{init} + T_{trf}^{\Delta w} + T_{con}^{\Delta w} = P_3 T_c. \qquad (A.2)$$

If we substitute Eq. (A.1) into Eq. (A.2), Eq. (A.2) can be rewritten as

$$P_1 T_c + T_{init} + T_{trf}^{\Delta w} + T_{con}^{\Delta w} = P_3 T_c. \qquad (A.3)$$

Likewise, between pipe processors i and $i+1$ where $3 \le i \le n-1$, we have,

$$P_1 T_c + (i-1)[T_{init} + T_{trf}^{\Delta w} + T_{con}^{\Delta w}] = P_{i+1} T_c. \qquad (A.4)$$

Since the sum of all the training patterns in all the processors must be equal to P_{tot}, we have,

$$P_1 + P_2 + \dots + P_n = P_{tot}.$$

Or,

$$P_1 T_c + P_2 T_c + \dots + P_n T_c = P_{tot} T_c. \qquad (A.5)$$

By letting $T = T_{init} + T_{trf}^{\Delta w} + T_{con}^{\Delta w}$ and substituting Eq. (A.1), (A.3) and (A.4) into Eq. (A.5), we have,

$$(P_1 T_c) + (P_1 T_c) + (P_1 T_c + T)... + (P_1 T_c + (n - 2)T) = P_{tot} T_c.$$

By grouping all the common terms together and simplifying, we have,

$$n P_1 T_c + \frac{T(n - 2)(n - 1)}{2} = P_{tot} T_c \qquad (A.6)$$

Rewriting Eq. (A.6), we can express $P_1 T_c$ in terms of $P_{tot} T_c$ and T. That is, we have,

$$P_1 T_c = \frac{2 P_{tot} T_c - T(n - 2)(n - 1)}{2n} \qquad (A.7)$$

Substituting Eq. (A.7) into Eq. (A.4) and letting $i = n - 1$ we have,

$$P_n T_c = \frac{2 P_{tot} T_c + T(n - 2)(n + 1)}{2n} \qquad (A.8)$$

Referring to the timing diagram of Fig. (A.1), we note that the time per epoch, T_{epoch} is the time taken by the last pipe processor, n. Summing all the timing components for pipe processor n, we have

$$T_{epoch} = T_{con}^{\Delta w_0} + P_n T_c + T + T_{con}^w + (n - 1)(T_{init} + T_{trf}^w)$$

The above expression can be written in the following form.

$$T_{epoch}^{P-ring} = T_{con}^{\Delta w_0} + T + T_{con}^{w} + (n-1)(T_{init} + T_{trf}^{w}) \quad (A.9)$$

$$+ \frac{2P_{tot}T_c + T(n-2)(n+1)}{2n}$$

A.2 Theoretical Optimal Epoch Time for Ring Topology

This section describes the time per epoch for the ring topology. The time per epoch for the theoretical optimal pattern distribution amongst a homogeneous array of processors is derived.

In order to obtain the time per epoch, we need to know the steps involved in executing one epoch of the parallelized backpropagation algorithm when the ring topology is used.

1. The weight changes and bias changes for the current epoch are initialized to zero at the start.

2. After the initialization, the forward pass and the backward pass of the backpropagation algorithm is performed for each training pattern assigned to the processor. The quadratic error between the desired and actual output is also calculated.

3. The total weight change and error due to all the patterns assigned to the processor are calculated by accumulating the weight change and error calculated for individual patterns in step 2.

4. An all-to-all broadcast of the weight and bias changes is performed. For this operation, each pipe processor will send its weight and bias changes and error to its downstream (right) neighbor while waiting to receive the weight and bias changes and error from the upstream neighbor. In a n transputer network, there will be $n - 1$ transfers of weight and bias changes and error. At the end of the transfers, each processor will receive all the weight and bias changes and errors from all the other processors.

5. Each processor will accumulate the weight and bias changes and errors it has obtained during the all-to-all broadcast. There will be a total of $N - 1$ accumulations as each processor does not need to add its own weight and bias changes.

6. Each processor will then update the weights as well as checking whether convergence is reached.

For the ring topology, it can be noticed that the flow of the data is only in one direction. For a homogeneous array of processors, the optimal pattern distribution occurs when equal distribution of patterns is chosen. This is because with equal distribution there will be no waiting time involved in each processor. That is, if there is a total of P_{tot} number of training pattern present, the number of patterns in each processor should be P_{tot}/N.

By following the steps involved in this parallel version of the backpropagation algorithm, a timing diagram as shown in Fig. (A.2) can be drawn.

To obtain the time per epoch, we need to accumulate all the

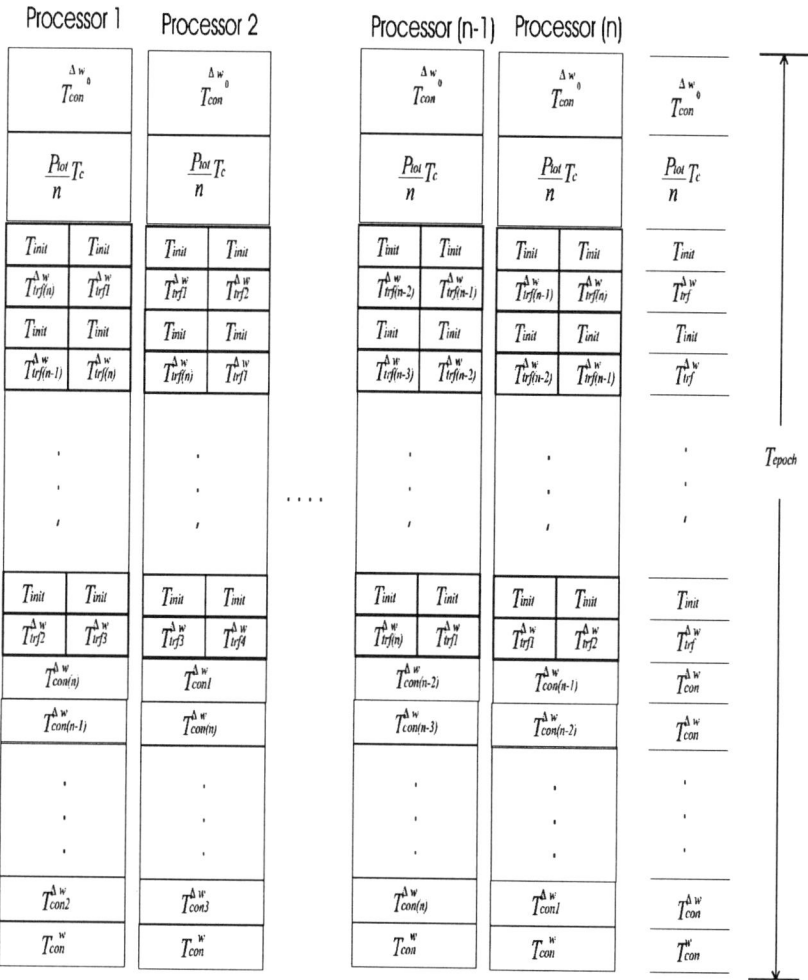

Figure A.2: Timing Diagram for an epoch for a homogeneous array of transputers with optimum pattern distribution for the ring topology.

timing components together. It can be seen clearly from Fig. (A.2) that the time per epoch is as follows:

$$T_{epoch}^{Ring} = T_{con}^{\Delta w_0} + (n-1)(T_{init} + T_{trf}^{\Delta w}) + (n-1)T_{con}^{\Delta w} + T_{con}^{w} \quad (A.10)$$

A.3 Comparison between Pipelined Ring Topology and Ring Topology

After obtaining the time per epoch for the theoretical optimum pattern distribution for the pipelined ring and ring topologies, we can find out under what conditions one topology is better than the other. This can be done by comparing their time per epoch equations. One way is to assume that pipelined ring outperforms the ring topology and examine when this is true. This assumption implies that the time per epoch in Eq. (A.10) is less than the time per epoch in Eq. (A.10). Since the number of weights and the number of weights increments are the same, we can set the time to transfer the weight changes and error as equal to the time taken to transfer the updated weights, i.e. $T_{trf}^{\Delta w} = T_{trf}^{w}$ in Eqs. (A.10) and (A.10).

Setting $T_{epoch}^{P-ring} < T_{epoch}^{Ring}$, we have,

$$(n-2)(n - \frac{T_{con}^{\Delta w} + T_{init} + T_{trf}^{\Delta w}}{T_{con}^{\Delta w} - T_{init} - T_{trf}^{\Delta w}}) > 0 \quad (A.11)$$

The above expression shows that for the optimized pipelined ring topology to produce a smaller epoch time than the optimized ring topology, two conditions must be satisfied. First, the number of processors used in the processor network must be greater than 2 $(n > 2)$ and the number of processors must also be greater than

$$\frac{T_{con}^{\Delta w} + T_{init} + T_{trf}^{\Delta w}}{T_{con}^{\Delta w} - T_{init} - T_{trf}^{\Delta w}}$$

i.e.

$$n > 2 \tag{A.12}$$

$$n > \frac{T_{con}^{\Delta w} + T_{init} + T_{trf}^{\Delta w}}{T_{con}^{\Delta w} - T_{init} - T_{trf}^{\Delta w}} \tag{A.13}$$

By substituting the elemental timings for the expressions $T_{con}^{\Delta w}$, T_{init} and $T_{trf}^{\Delta w}$, the term $\frac{T_{con}^{\Delta w} + T_{init} + T_{trf}^{\Delta w}}{T_{con}^{\Delta w} - T_{init} - T_{trf}^{\Delta w}}$ can be written in the form

$$\frac{T_{con}^{\Delta w} + T_{init} + T_{trf}^{\Delta w}}{T_{con}^{\Delta w} - T_{init} - T_{trf}^{\Delta w}} = \frac{t_{aas} + rd + \frac{t_{aas}}{N_{wb}}}{t_{aas} - rd + \frac{t_{aas}}{N_{wb}}}$$

Since the number of weights and biases, N_{wb}, is always many order of magnitude higher than t_{aas}, the term $\frac{t_{aas}}{N_{wb}}$ will be very small. The inequality Eq. (A.11) can then be simplified to

$$(n - 2)(n - \frac{t_{aas} + rd}{t_{aas} - rd}) > 0 \tag{A.14}$$

Substituting for t_{aas} and rd for the processors, we can find the value of n above which the pipelined ring will outperform the ring. For the T805-20 and T805-25 models of transputers, we substituted t_{aas} and rd and found n as 2.15 and 2.67 respectively. This implies that as long as there are three or more processors in the processor array, pipelined ring will outperform the ring topology in a transputer network.

Appendix B

A Sample Parallel C Program

This appendix provides a sample program listing written in 3L Parallel C. Together with the program listing, a configuration file is also included. This sample program waits for the user to key in two words, concatenate the two words together, change the concatenated word to upper case and then display the results on the screen. Three tasks are involved in this program. They are the *master* task, the *concat* task and the *upper* task. The *master* task is the administrative task that obtains the two words from the user, pass the two words two the other two tasks for processing before receiving them back for display. The *concat* task takes two words, concatenate them and passes it down to the next task for further processing. And finally, the *upper* task converts a string from lower case to upper case and then pass it back to the *master* task. These three tasks are connected in a ring fashion where the *master* task heads the ring which is also connected to the host processor via the *filter* task and *afserver* program. Fig. (B.1) shows the configuration of the three tasks. These three tasks are placed within a single transputer.

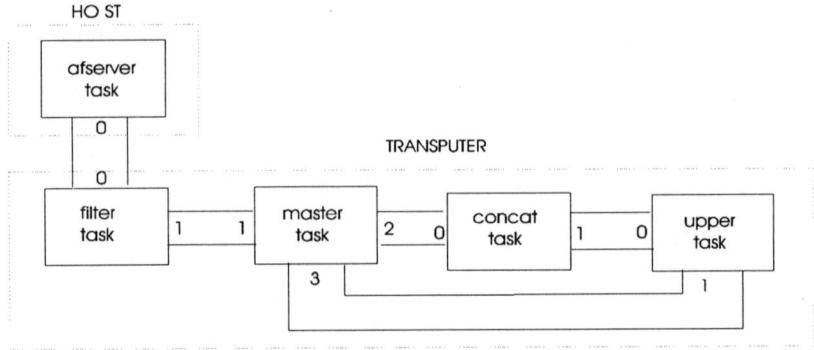

Figure B.1: Tasks allocations for sample program

The program segment below shows the configuration file needed by the various 3L configuration utilities. It serves the purpose to allow easy description of both the processor networks and how the tasks are being interconnected to one another to form the whole program structure.

```
!
! SAMPLE.CFG        configuration file for sample
!
! Hardware is made up of one PC and one transputer
!
processor host
processor root
wire ? host[0] root[0]
!
! Task declarations indicating channel I/O ports and
! memory requirements
!
```

```
! Sample shows that the afserver task has 1 port
! filter task has 2 ports with 10k of data memory
! master task has 4 ports with 10k of data memory
! concat task has 2 ports with 10k of data memory
! upper task has 2 ports with 10k of data memory
!
task afserver ins=1 outs=1
task filter ins=2 outs=2 data=10K
task master ins=4 outs=4 data=10k
task concat ins=2 outs=2 data=10k
task upper ins=2 outs=2 data=10k
!
! Assign software tasks to physical processors
!
! Sample shows that all tasks are placed
! in the transputer except for the afserver task
! which is placed in the host processor
!
place afserver host
place master root
place concat root
place upper root
place filter root
!
! Set up the connections between the tasks
!
connect ? afserver[0] filter[0]
connect ? filter[0] afserver[0]

connect ? filter[1] master[1]
```

```
connect ? master[1] filter[1]

connect ? master[2] concat[0]
connect ? concat[0] master[2]

connect ? concat[1] upper[0]
connect ? upper[0] concat[1]

connect ? upper[1] master[3]
connect ? master[3] upper[1]
```

The program segment below shows the program listing for the Parallel C program of the main task, *master.c*.

```
/*master.c*/
/*main routine to get two words from the keyboard,
and sends them over to the other tasks for processing
before retrieving them back for display*/

#include <chan.h>
#include <stdio.h>

main(argc, argv, envp, in_ports, ins, out_ports, outs)
int argc, ins, outs;
char *argv[], *envp[];
CHAN *in_ports[], *out_ports[];
{
    int length;
    char str1[50];
    char str2[50];
    char str3[100];
    for (;;) {

    /*obtain the first word*/
    printf("\nKey in first word: ");
    gets(str1);

    /*obtain the second word*/
    printf("\nKey in second word: ");
    gets(str2);
```

```
    /*sends the first word across the channel*/
    chan_out_word(strlen(str1), out_ports[2]);
    chan_out_message(strlen(str1),str1, out_ports[2] );

    /*sends the second word across the channel*/
    chan_out_word(strlen(str2), out_ports[2]);
    chan_out_message(strlen(str2),str2, out_ports[2] );

    /*receives the concatenated words from the channel*/
    chan_in_word(&length, in_ports[3]);
    chan_in_message(length, str3, in_ports[3] );

    /*displays the processed words*/
    puts(str3);
        }
}
```

The program segment below shows the program listing for the
Parallel C program of the task, *concat.c.*

```c
/*concat.c*/
/*routine to concatenate two strings together*/

#include <chan.h>

main(argc, argv, envp, in_ports, ins, out_ports, outs)
int argc, ins, outs;
char *argv[], *envp[];
CHAN *in_ports[], *out_ports[];
{
    int  i,j;
    int  length1,length2,length3;
    char str1[50];
    char str2[50];
    char str3[100];

    for (;;) {

    /*receives the first word from the main task*/
    chan_in_word(&length1, in_ports[0]);
    chan_in_message(length1, str1, in_ports[0]);

    /*receives the second word from the main task*/
    chan_in_word(&length2, in_ports[0]);
    chan_in_message(length2, str2, in_ports[0]);

    /*copies the first word to destination, str3*/
```

```
for(i=0;i<length1;i++)
str3[i]=str1[i];

/*insert a space character to separate two words*/
str3[i]=' ';

/*copies the second word to destination, str3*/
for(j=0,i=length1+1;j<length2;j++,i++)
str3[i]=str2[j];

/*add a NULL character to the end of string*/
str3[i]=0;

/*specifies the length of the resultant string*/
length3=length1+length2+2;

/*sends the concatenated words over the channel*/
chan_out_word( length3, out_ports[1] );
chan_out_message( length3, str3, out_ports[1] );
        }
}
```

The program segment below shows the program listing for the Parallel C program of the task, *upper.c*.

```c
/*upper.c*/
/*routine to receive a string from a upstream task,
converts that string to upper case before sending it
down to the next downstream task*/
#include <chan.h>
#include <ctype.h>

main(argc, argv, envp, in_ports, ins, out_ports, outs)
int argc, ins, outs;
char *argv[], *envp[];
CHAN *in_ports[], *out_ports[];
{
    int length,i;
    char str[100];

    for (;;) {

    /*receives the string from upstream task*/
    chan_in_word(&length, in_ports[0]);
    chan_in_message(length, str, in_ports[0]);

    /*converts string to upper case*/
    for(i=0;i<length;i++)
    str[i]=toupper(str[i]);

    /*sends the converted string to next task*/
    chan_out_word( length, out_ports[1] );
```

```
chan_out_message( length, str, out_ports[1]);
        }
}
```

The listing below shows a batch file written for MS-DOS version 6.2 operating system that will compile and run the sample program given.

```
REM sample.bat
REM Compile, link, and configure the sample program
REM Compiling the tasks
c:\tc2v2\t8c master
c:\tc2v2\t8c concat
c:\tc2v2\t8c upper

REM Linking the tasks
command/c c:\tc2v2\t8ctask master
command/c c:\tc2v2\t8cstask concat
command/c c:\tc2v2\t8cstask upper

REM Configuring the tasks together
c:\tc2v2\config  sample.cfg sample.app

REM Running the compiled application
c:\tc2v2\afserver -:b sample.app
```

Appendix C

The branch and bound method for solving mixed integer programming problems

This appendix outlines the branch and bound method of solving the mixed integer programming (MIP) [72]. A mixed integer programming problem can be expressed in the following form.

$$Maximize \quad z = CX + DY \qquad (C.1)$$
$$X, Y$$

$$Subject \ to$$

$$AX + DY \leq b \qquad (C.2)$$

$$X \geq 0, Y \geq 0 \qquad (C.3)$$

$$and \quad X \ is \ integer \qquad (C.4)$$

The objective of the mixed integer programming problem is to maximize the cost function z which is a linear function of X and Y. In the above equations, C and D are row vectors of dimension $n \times 1$

and $n' \times 1$ respectively. X and Y are column vectors of dimension $1 \times n$ and $1 \times n'$ respectively. X and Y are variables that are to be determined so that their contribution will maximize z. A is a m by n matrix while D is a m by n' matrix. Both A and D are the constraint matrices that both variables X and Y must satisfy.

In the problem formulation above, if only Eq. (C.1) to (C.3) are considered, the problem is reduced to the linear programming problem which can be solved readily with the Simplex method [55]. This problem is denoted by $LP0$. For this problem, we notice that the integer constraints for X is not imposed. The values of X are considered free and can take fractional values. Once the integer constraint is added, X is no longer free. The branch and bound method will constraint values of X that have fractional components so that only integer values are considered. The steps of the branch and bound approach are outlined below.

1. (Initialization) Set z^*, the current best solution to the MIP, to some arbitrary small value. Denote the initial problem with X free as $LP0$. Solve this linear programming problem. This will be considered the first subproblem (node) of the branch and bound tree. If $LP0$ is infeasible, then the MIP is also infeasible. Hence the algorithm terminates. If the solution for $LP0$ has integer solutions for all values of X, then the optimal solution for the MIP is found and the algorithm terminates. Otherwise, there will be at least 1 or more elements of X that are fractional. In this case, proceed to step 2.

2. (Branching) From the optimal linear programming solution, pick a variable x_k from X that does not satisfy the integer constraint. Let $x_k = t$ where t is a non integral number. It

is noted that since x_k must be an integer and it is impossible to find an integer in the range $\lfloor t \rfloor < x_k < \lfloor t \rfloor + 1$, x_k can only be found in the range $x_k \leq \lfloor t \rfloor$ and $x_k \geq \lfloor t \rfloor + 1$. Hence, we create two new subproblems (nodes) by introducing the inequality $x_k \leq \lfloor t \rfloor$ to the first subproblem (node) and the inequality $x_k \geq \lfloor t \rfloor + 1$ to the second subproblem (node). Solve the subproblems at each of these nodes. For nodes that have infeasible solutions or have the objective solutions smaller than z^*, drop them from further consideration. Check each dangling node to see if it satisfies all the constraints for the MIP. If so, record it down. Compare the solutions of these nodes to z^*. Keep nodes that have objective values greater than z^* in a list and label them as dangling. Replace z^* with the highest objective value obtained so far. This new value of z^* will become the upper bound for the MIP problem thus far.

3. (Termination Test) If the dangling node list is empty, check to see whether any feasible MIP solution has been recorded. If so, the optimal MIP has been found. If not, no solution exists and the algorithm terminates.

4. (Bounding) If the dangling node list is not empty, consider the most recently created subproblem for further branching (step 2). This will lead us down one side of the branch and bound tree and quickly finds a candidate solution. Then we backtrack our way up to the top of the other side of the tree.

To illustrative the branch and bound method of solving MIP problem, consider the following problem.

$Maximize \quad z = 100x_1 + 20x_2 + 100x_3 + 30y_1 + 15y_2$

Subject to

$$3x_1 + 4x_2 + 25x_3 + 50y_1 + 3y_2 \leq 100$$

$$-88x_1 + 30x_2 + 3x_3 + 5y_1 - 10y_2 \leq 70$$

$$16x_1 + 2x_2 + x_3 - 5y_1 + 12y_2 \leq 30$$

$$x_1, x_2, x_3 \geq 0, y_1, y_2 \geq 0$$

and $\quad\quad\quad x_1, x_2, x_3 are integers$

The steps for for the branch and bound method is shown below.

1. (LP0) Set $z^* = 0$. Next solve the initial problem with x_1, x_2 and x_3 free, the relaxed solution is

$$z = 544.08, x_1 = 1.63, x_2 = 0, x_3 = 3.80, y_1 = 0, y_2 = 0$$

This is the best solution that can be obtained for the linear programming problem. Any solution obtained with the integer constraints included will be lower than this. It is noted that in this solution set, x_1 and x_3 are non integral. Choosing arbitrary x_1, we create two subproblems by imposing the constraints $x_1 \leq 1$ (LP1) and $x_1 \geq 2$ (LP2). Let us consider LP1.

2. (LP1) The solution for this with the constraint $x_1 \leq 1$ added is

$$z = 508.18, x_1 = 1, x_2 = 4.99, x_3 = 3.07, y_1 = 0, y_2 = 0.08$$

Notice that the value of z obtained now is lower than that obtained in LP0. In this solution set, x_2 and x_3 are non integral. We select x_2 randomly for further consideration. Two additional subproblems obtained are LP3 and LP4. The new constraints added are $x_2 \leq 4$ and $x_2 \geq 5$.

3. (LP3) Considering LP3, the solution set is

$$z = 504.70, x_1 = 1, x_2 = 4, x_3 = 3.21, y_1 = 0, y_2 = 0.23$$

x_3 is still non integral. We introduce two more subproblems LP5 and LP6 by imposing the constraints $x_3 \leq 3$ and $x_3 \geq 4$.

4. (LP5) If we solve for LP5, we have

$$z = 487.46, x_1 = 1, x_2 = 4, x_3 = 3, y_1 = 0.10, y_2 = 0.29$$

Since x_1, x_2, x_3 are all integer values. A solution for the MIP is found. Since the value of z is higher than z^*, we replace z^* with $z^* = 487.46$. This is the upper bound for the MIP. Any subproblem considered later that has a value of z value lower than z^* will be discarded.

5. (LP6) We now backtrack and find that LP6 is not yet solved. Solving LP6, we have

$$z = 400.00, x_1 = 0, x_2 = 0, x_3 = 4, y_1 = 0, y_2 = 0$$

Since the value of z obtained is lower than z^*, this node is discarded for further consideration.

6. (LP4) We backtrack further up the tree and we find that there is a dangling node at node 4. Solving this, we have

$$z = 497.86, x_1 = 1, x_2 = 5, x_3 = 2.94, y_1 = 0.06, y_2 = 0.11$$

Since x_3 is non integral and z is higher than z^*, node 4 is a dangling node. We now create two new subproblems LP7 and LP8 by introducing two new constraints $x_3 \leq 2$ and $x_3 \geq 3$.

7. (LP7) Solving for LP7, we have

$$z = 422.78, x_1 = 1, x_2 = 5.10, x_3 = 2, y_1 = 0.51, y_2 = 0.36$$

Since the value of z is lower than z^*, this node is discarded for any further consideration.

8. (LP8) We now consider LP8. We find that there is no feasible solution for this subproblem. Hence it is discarded for any further consideration.

9. (LP2) By backtracking, we find that the only dangling node left is node 2. Solving, we have

$$z = 436.11, x_1 = 2, x_2 = 0, x_3 = 2.11, y_1 = 0.82, y_2 = 0$$

Since the value of z is lower than z^*, we discard this node for any consideration.

10. We notice that all dangling nodes have been investigated. The best MIP solution found so far is at node 5. Thus, the optimal solution for this MIP is

$$z = 487.46, x_1 = 1, x_2 = 4, x_3 = 3, y_1 = 0.10, y_2 = 0.29$$

An illustration for the branch and bound method is shown in Fig. (C.1).

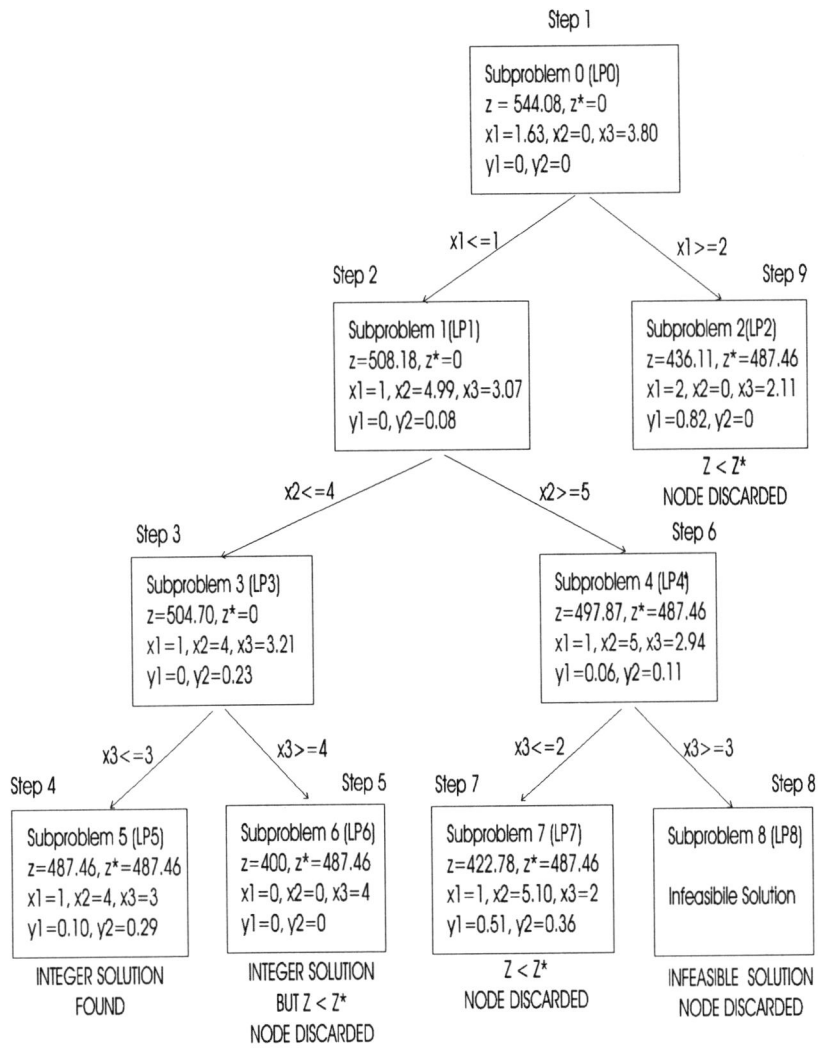

Figure C.1: Branch and bound tree for the sample problem

Bibliography

[1] Thomas Miller, Richard Sutton and Paul Werbos, *Neural Networks for Control.* MIT Press, Cambridge, Mass., 1990.

[2] R.Linggard, D.J.Myers and C.Nightingale, *Neural Networks for Vision Speech and Natural Language.* Chapman and Hall, London, UK, 1992.

[3] Michael Azoff, *Neural Network Time Series Forcasting of Financial Markets.* John Wiley & Sons, Chichester, England, 1994.

[4] Ben Yuhas and Nirwan Ansari, *Neural Networks in Telecommunications.* Kluwer Acadamic Publishers, Boston, 1994.

[5] Malcom Taylor and Paulo Lisboa, *Techniques and Applications of Neural Networks.* Ellis Horwood, Chichester, England, 1993.

[6] D.E. Rumelhart and G.E. Hinton and R.J.,Williams, "Learning internal representations by error propagation," *Nature*, vol. 323, pp. 533–536, 1986.

[7] D.E. Rumelhart and J.L. McClelland, *Parallel Distributed Processing: Explorations in the Microstructure of Cognition: Foundations*, vol. 1, pp. 318–362. MIT Press, Cambridge, Mass., 1986.

[8] Y. Le Cun, B.Boser, J.S.Denker, D.Henderson, R.E.Howard, W.Hubbard and L.D.Jackel, "Backpropagation applied to handwritten zip code recognition," *Neural Computation*, vol. 1, no. 4, pp. 541–551, 1989.

[9] A.Waibel, "Consonant recognition by modular construction of large phonetic time-delay neural networks," *Advances in Neural Information Processing Systems* (D.Touretzkey, ed.), pp. 215–223, Morgan Kaufman, San Mateo, CA., 1989.

[10] R.Jacobs, "Increased rates of convergence through learning rate adaptation," *Neural Networks*, vol. 1, no. 4, pp. 295–308, 1988.

[11] S.Kolias and D.Anastassiou, "An adaptive least squares algorithm for the efficient training of artificial neural networks," *IEEE Transactions on Circuits and Systems*, vol. 36, pp. 1092–1101, 1989.

[12] Y.LeCun, "Generalization and network design strategies," *Connectionism in Perspective*, pp. 143–155, North Holland, Amesterdam, 1989.

[13] R.Battiti, "First- and second-order methods for learning: Between steepest descent and newton's method," *Neural Computation*, vol. 4, no. 2, pp. 141–166, 1992.

[14] Harry A.C. Eaton and Tracy L.Olivier, "Learning coefficient dependence on training set size," *Neural Networks*, vol. 5, pp. 283–288, 1992.

[15] S.W. Aiken, Mark W.Koch and Morian W.Roberts, "A parallel neural network simulator," *Proceedings of the ICNN International Conference on Neural Networks*, vol. 2, (San Diego, California), pp. 611–616, 1990.

[16] F.Baiardi, R.Mussardo, R.Serra and G. Valastro, "Feedforward layered networks on message passing parallel computers," *Second Italian Workshop on Parallel Architectures and Neural Networks*, pp. 213–222, World Scientific Publishing Co, 1990.

[17] Urs A.Muller, Anton Gunzinger and Walter Guggenbuhl, "Fast neural net simulation with a DSP processor array," *IEEE Transactions on Neural Networks*, vol. 6, no. 1, pp. 203–213, 1995.

[18] Soheif Shams and Jean-Luc Gaudiot, "Implementing regularly structured neural networks on the DREAM machine," *IEEE Transactions on Neural Networks*, vol. 6, no. 2, pp. 407–421, 1995.

[19] Helene Paugam-Moisy, "Parallel neural computing based on network duplicating," *Parallel Algorithms for Digital Image Processing Computer Vision and Neural Networks* (Ioannis Pitas, ed.), pp. 305–340, John Wiley & Sons Ltd., 1993.

[20] Alexander Singer, "Implementations of artificial neural networks on the Connection Machine," *Parallel Computing*, vol. 14, pp. 305–315, May 1990.

[21] G. Richards, "Implementation of backpropagation on a trans-puter array," *Developments Using Occam* (J. Kerridge, ed.), pp. 173–179, IOS Press, Amesterdam, 1988.

[22] M.Forti, "A neural network for signal decomposition prob-lems," *International Journal of Circuit Theory and Applica-tions*, vol. 19, pp. 65–75, 1991.

[23] A.Cichocki and R.Unbehauen, *Neural Networks for Optimiza-tion and Signal Processing.* John Wiley & Sons, 1993.

[24] A. Petrowski and Helene Paugam-Moisy, "Parallel neural com-putation based on algebraic partitioning," *Parallel Algorithms for Digital Image Processing Computer Vision and Neural Net-works* (Ioannis Pitas, ed.), pp. 259–304, John Wiley & Sons Ltd., 1993.

[25] S.Y. Kung and Wei-Hsin Chou, "Mapping of neural networks onto VLSI array processors," *Parallel Digital Implementa-tion of Neural Networks* (K.Wojtek Przytula and Viktor K. Prasanna, eds.), pp. 3–49, Prentice Hall International Inc, En-glewood Cliffs, New Jersey, 1993.

[26] J. del R. Millan and P.Bofill, "Learning by backpropagation: a systolic algorithm and its transputer implementation," *In-ternational Journal of Neural Networks - Research and Appli-cations*, vol. 1, no. 3, pp. 119–137, 1989.

[27] Hyunsoo Yoon, Jong H. Nang and S.R.Maeng, "Parallel simulation of multilayered neural networks on distributed-memory multiprocessors," *Microprocessing and Micropro-gramming*, vol. 29, pp. 185–195, July 1990.

[28] Vipin Kumar, Shashi Shekar and Minesh B.Amin, "A scalable parallel formulation of the backpropagation algorithm for hypercubes and related architectures," *IEEE Transactions on Parallel and Distributed Systems*, vol. 5, no. 10, pp. 1073–1090, 1994.

[29] Arularasan, R., P. Saratchandran and N. Sundararajan "A study on network partitioning of fully connected backpropagation neural networks with equal distribution of nodes," EEE/CSP/9503, Nanyang Technological University, Nanyang Avenue, Singapore, July 1995.

[30] Shou King Foo, P. Saratchandran and N. Sundararajan, "Comparison of pattern allocation schemes for parallel implementation of backpropagation neural networks," *Proceedings of the Third International Conference on Automation, Robotics and Computer Vision 94, ICARCV 94*, vol. 1, (Singapore), pp. 575–579, November 1994.

[31] Shou King Foo, P. Saratchandran and N. Sundararajan, "A theoretical study of training set parallelism for backpropagation networks on a transputer array," *Proceedings of World Congress on Neural Networks 94, WCCN 94*, vol. 2, (San Diego, U.S.A.), pp. 519–524, June 1994.

[32] Shou King Foo, P. Saratchandran and N. Sundararajan, "Parallel implementation of backpropagation on transputers," *Proceedings of 1993 International Joint Conference on Neural Networks,IJCNN 93*, vol. 3, (Nagoya, Japan), pp. 3058–3061, October 1993.

[33] Hyunsoo Yoon, Jong H. Nang and S.R.Maeng, "Neural networks on parallel computers," *Neural and Intelligent Systems Integration* (Branko Soucek and the IRIS group, eds.), pp. 235–279, John Wiley & Sons Ltd., 1991.

[34] Dean A. Pomerleau, George L.Gusciora, David Touretzky and H.T.Kung, "Neural network simulation at WARP speed: How we got 17 million connections per second," *IEEE International Conference on Neural Networks*, vol. 2, (San Diego, California), pp. 143–150, July 1988.

[35] X.Liu and G.L.Wilcox, "Benchmarking of the CM-5 and the cray machines with a very large backpropagation neural network," *Proceedings of IEEE International Conference on Neural Networks*, (Orlando Florida), pp. 22–27, July 1994.

[36] H.Kato, H.Yoshizawa, Hiroki Ichiki and K.Asakawa, "A parallel neurocomputer architecture towards billion connection updates per second," *Proceedings of the International Joint Conference Neural Networks (IJCNN-90)*, vol. 2, (Washington D.C), pp. 47–50, October 1990.

[37] H.Yoshizawa, H.Kato , Hiroki Ichiki and K.Asakawa, "A highly parallel architecture for backpropagation using a ring register data path," *Proceedings of the Second International Conference Microelectronics and Neural Networks (ICMNN-91)*, (Munich), pp. 325–332, October 1991.

[38] V.Benes, "Optimal rearrangable multistage connecting networks," *Bell System Technical Journal*, vol. 43, pp. 1641–1656, 1964.

[39] M.Witbrock and M.Zagha, "Backpropagation learning on the IBM GF11," *Parallel Digital Implementation of Neural Networks* (K.Wojtek Przytula and Viktor K. Prasanna, eds.), pp. 77–104, Prentice Hall International Inc., Englewood Cliffs, New Jersey, 1993.

[40] Yoshiji Fujimoto,N.Fukuda and T.Akabane, "Massively parallel architectures for large scale neural network simulations," *IEEE Transactions on Neural Networks*, vol. 3, pp. 876–888, November 1992.

[41] David May, Roger Shepherd and Peter Thomson, "Next generation transputers," *Transputers'92* (M.Becker, L.Litzler and M.Trehel, eds.), pp. 345–351, IOS Press, Oxford, UK, 1992.

[42] U.de Carlini and U.Villano, *Transputers and Parallel Architectures*. Ellis Horwood Limited, Chichester, England, 1991.

[43] Ian Graham and Tim King, *The Transputer Handbook*. Prentice Hall, 1990.

[44] D.A.P.Mitchel, J.A.Thomson, G.A.Manson and G.R.Brooks, *Inside The Transputer*. Blackwell Scientific Publications, Oxford, England, 1990.

[45] Russell Eberhart and Roy W. Dobbins, *Neural Networks PC Tools: A Practical Guide*. Academic Press, San Diego, California, 1990.

[46] Weidong Pan and Peter K. Sharpe, "A simulation of multilayer perceptrons exploiting training set parallelism," *Proceedings of International Joint Conference of Neural Networks*, vol. 1, (Beijing, China), pp. 183–188, November 1992.

[47] 3 L Ltd, *Parallel C Version 2.2.2 Release Notes*, 1991.

[48] Fahlman, S.E., "An empirical study of learning speed in backpropagation networks," CMU-CS-88-162, Carnegie Mellon University, School of Computer Science, Pittsburgh, USA, 1988.

[49] Sejnowski,T.J. and Rosenberg, C.R., "Parallel networks that learn to pronounce english text," *Complex Systems*, vol. 1, pp. 145–168, 1987.

[50] Gorman, R.P. and Sejnowski, T.J., "Analysis of hidden units in a layered network trained to classify sonar targets," *Neural Networks*, vol. 1, pp. 75–89, 1988.

[51] Shou King Foo, P. Saratchandran and N. Sundararajan, "Analysis of training set parallelism for backpropagation neural networks," *International Journal of Neural Systems*, vol. 6, no. 1, pp. 61–78, 1995.

[52] Shou King Foo, P. Saratchandran and N. Sundararajan, "A study of parallel implementation of backpropagation neural networks on a transputer array," *Neural, Parallel and Scientific Computations*, vol. 2, pp. 321–332, 1994.

[53] K. Wojtek Przytula and Viktor K. Prasanna, *Parallel Digital Implementations of Neural Networks*. PTR Prentice Hall, Englewood Cliffs, NJ, 1993.

[54] Shou King Foo, P. Saratchandran and N. Sundararajan, "Comparison of parallel and serial implementation of feedforward neural networks," *Journal of Microcomputer Applications*, vol. 17, pp. 83–94, 1995.

[55] Harvey M. Salkin and Kamlesh Mathur, *Foundations of Integer Programming*. North-Holland, New York, 1989.

[56] Numerical Analysis Group, Oxford, *NAG FORTRAN library, MARK 14*, 1990.

[57] Shou King Foo, P. Saratchandran and N. Sundararajan, "Parallel implementation of backpropagation neural networks on a heterogenous array of transputers," *to appear inIEEE Transactions on Systems, Man, and Cybernetics*, 1997.

[58] A.Barto and M.Duff, "Montecarlo matrix inversion and reinforcement learning," *Advances in Neural Information Processing Systems*, pp. 655–662, 1994.

[59] Winston, Wayne L., *Operations research: applications and algorithms*. PWS-KENT Publishing Company, second ed., 1991.

[60] John H. Holland, *Adaption in Natural and Artificial Systems*. The University of Michigan Press, Ann Arbor, 1975.

[61] Lawerence Davis, *Handbook of Genetic Algorithms*. Van Nostrand Reinhold, New York, 1991.

[62] Kalyanmoy Deb and David E. Goldberg, "A comparative analysis of selection schemes used in genetic algorithms," *Foundations of Genetic Algorithms*, pp. 69–93, Morgan Kaufamm, San Mateo, California, 1991.

[63] Shou King Foo, P. Saratchandran and N. Sundararajan, "Genetic algorithm based pattern allocation schemes for parallel implementations of backpropagation neural networks," EEE/CSP/9502, Nanyang Technological University, Nanyang Avenue, Singapore, May 1995.

[64] David E. Goldberg, *Genetic Algorithms in Search, Optimization, and Machine Learning*. Addison-Wesley, Reading, Massachusetts, 1989.

[65] Shou King Foo, P. Saratchandran and N. Sundararajan, "Genetic algorithms based pattern allocation schemes for training set parallelism in backpropagation neural networks," *Proceedings of IEEE International Conference on Evolutionary Computing 95, ICEC 94*, (Perth, Western Australia), Nov 1995.

[66] Shou King Foo, P. Saratchandran and N. Sundararajan, "Applications of genetic algorithm for parallel implementation of backpropagation neural networks," *Proceedings of International Symposium on Intelligent Robotic Systems, ISIRS 95*, (Bangalore, India), Nov 1995.

[67] William H. Press et. al., *Numerical Recipes in C*. Cambridge University Press, Cambridge, 1989.

[68] Srinivas, M. and Lalit M. Patnaik, "Genetic algorithms: A survey," *IEEE Computer*, pp. 17–26, 1994.

[69] Jose L. Ribeiro Filho and Philip C. Treleaven, "Genetic-algorithm programming environments," *IEEE Computer*, pp. 28–43, 1994.

[70] David Beasley, David R. Bull and Ralph R. Martin, "An overview of genetic algorithms: Part 1, fundamentals," *University Computing*, vol. 15, no. 2, pp. 58–69, 1993.

[71] Goldberg, D.E., Kargupta, H., Horn, J. and Cantu-Paz, E., "Critical deme size for serial and parallel genetic algorithms,"

illigal report 95002, The Illinois GA Lab, University of Illinois, 1995.

[72] Dakin, R.J., "A tree search algorithm for mixed integer programming problems," *Computer Journal*, vol. 8, pp. 250–255, 1965.

Index